CW01261316

Design Your Own Yacht

Ben Smith C.Eng., M.Sc., M.R.I.N.A.

ADLARD COLES
8 Grafton Street, London W1

Adlard Coles
William Collins Sons & Co. Ltd
8 Grafton Street, London W1X 3LA

First published in Great Britain by
Adlard Coles 1988

Copyright © Ben Smith 1988

British Library Cataloguing in Publication Data

Smith, B. (Ben)
  Design your own yacht.
  1. Yacht-building
  I. Title
  623.8′1223   VM331

ISBN 0-229-11760-0

Typeset by Columns of Reading
Printed and bound in Great Britain by
Butler & Tanner Ltd, Frome, Somerset

All rights reserved. No part of this publication may be reproduced, stored in a retrieval system, or transmitted, in any form, or by any means, electronic, mechanical, photocopying, recording or otherwise, without the prior permission of the publisher.

Every care has been taken to ensure that the information and instructions contained in this book are accurate. However, the author, editors and publisher accept no liability whatsoever for accidents or losses of any kind occasioned by reference to, or arising from information contained in this book.

CONTENTS

PREFACE		v
ACKNOWLEDGEMENTS		vii
CHARTING A COURSE		viii
1	**SETTING SAIL** *An Introduction*	1
2	**SALTY TALK** *An explanation of design terms*	7
3	**BLOCK AND TACKLE OF DESIGN** *Instruments for drawing the design*	17
4	**CHARIOTS OF SEA AND SAIL** *Types of yachts*	25
5	**TIMBERS THAT KISS THE WATERS** *Materials of yacht design*	30
6	**UNDER CANVAS** *Sails and rigging*	35
7	**SEA HORSES** *Selection of an engine*	41
8	**FITTINGS AND ABLE EQUIPMENT** *Discussing the outfit and fittings of a yacht*	46
9	**LOG OF DESIGN** *The specification*	51
10	**A SPIRAL ASPIRATION** *Design procedures*	56
11	**AHEAD AND ASTERN – ABOVE AND BELOW** *The contours of design*	62

Contents

12	**NAVIGATING THE CALCULATIONS** *Basic calculations of design*	68
13	**AS THE WATERS SLIP BY** *A brief explanation of flow and resistance round the hull design*	80
14	**SHAPING UP TO THE SEAS** *Finding the form characteristics of design*	88
15	**DRIVING WINDS** *Calculation and design of sails*	98
16	**POWER AND PROP** *Estimate of horsepower and propeller for design*	103
17	**OF SHOES AND SHIPS AND SEALING WAX** *Drawing the general arrangement for the design*	114
18	**A DESIGN HAND AT THE TILLER** *Design of rudder and steering*	127
19	**SHIPSHAPE AND BRISTOL FASHION** *Drawing the Lines Plan*	134
20	**BUOYANT AND AWEIGH** *Calculation for displacement, weight, and centres of the design*	141
21	**A DRUNKEN DESIGN AT SEA** *The stability of a yacht*	153
22	**HEARTS OF OAK** *The strength of a yacht*	163
23	**TIMBERS THAT DO NOT SHIVER** *Construction Plans*	171
24	**SLOP CHEST** *Ancillary elements of yacht design*	181
25	**TIED UP AT THE PORT OF DESIGN** *Final calculations*	185
26	**INTO DEEPER DESIGN WATERS** *More sophisticated elements of yacht design*	190
INDEX		**200**

PREFACE

The many yachtsmen and women I have met have been individuals in the true sense of the word. Eccentric at times – but always interesting.

Maybe it's because sailing offers freedom, challenge and adventure, demands physical and mental abilities of a high order, and requires a capacity to laugh at adverse winds or wryly smile when the storm clouds gather.

Yachtsmen play and battle with the winds of nature, lie close to its shores and waters and breathe its clean air. They forge friendships in a calm and eat and drink with hearty digestion. No wonder increasing numbers of characters are seeking sail and sea in order to break away from the restrictions of present-day society.

But it goes further. Yachtsmen have an *affinity* with their craft. Many maintain – or even build – their own vessels. It's not such a great step to even design one's own yacht – which is the claim of this book.

Do-it-yourself (DIY) books now form a prominent part of non-fiction literature. They probably fulfil our desire for self-creation, which is ever being frustrated by such distractions as TV. Numbers of texts exist which instruct on how to build one's own yacht or craft, but none – as far as I am aware – that shows the steps of design, except erudite works written for the professional yacht designer.

A word about this book. Communicators and educators consistently stress that knowledge should be imparted in simple and concise terms. The trouble is, very few follow this principle. Academics and professionals are reluctant to de-mystify their expertise and thus lose status. Simplicity itself is no guarantee that the recipient or learner will respond. It can often appear banal, impersonal and, frankly, boring. Style, imagery and a sense of reality are necessary adjuncts to interesting and informative digests or lectures.

I have purposely aimed to achieve such goals.

The book covers not only the rudiments of yacht design in clear and simple fashion (with minimal demands on mathematical ability) it also strives towards an entertaining approach. The chapters include some historical background on sailing vessels and naval architecture, autobiographical sketches and humorous anecdotes related to the subject, together with artistic and easily interpretable illustrations.

It can be read as a tool or aid to design, or merely for pleasure and a versed understanding of one's special interest – yachts and yachting. It should also prove useful to students and practitioners of yacht and small craft design.

Now read on – and good luck with your design!

ACKNOWLEDGEMENTS

Acknowledgements are given to the following organisations for assistance in providing information and/or illustrations for the book.

Lloyd's Register of Shipping for use of part of their rules, tables and illustrations from their Rules and Regulations for the Classification of Yachts and Small Craft.

Perkins Engines Ltd for use of information, tables and illustrations from their installation and marine power manuals.

Cougar Marine Ltd for photographs of their marine constructions, including the British challenger for the America's Cup; Racal-Decca, and Sperry Marine for their navigational aids; British Maritime Technology, Suzuki, Southern Ocean Shipyard, N.B. Yachts, Ltd, and other companies for various illustrations of their products.

Acknowledgement also to the illustrator, John Barfoot, who additionally provided technical back-up to the material in the book.

Artistic illustrations by the author. Technical illustrations by John Barfoot.

> This book is dedicated to my children
> Pamela and Mark who, since the death
> of their mother, have given staunch
> support to my writing endeavours.

CHARTING A COURSE

The shortest distance between conception and creation of an idea is the straightforward approach. This book adopts that principle, charting a straight course in explaining the steps of yacht design. Occasionally – as will be appreciated by you yachtsmen readers – I have had to tack in order to make headway, diversifying into explanation of certain calculations or aspects of design before bringing them into the total framework. But in general, I have steered a straight course.

Each chapter leads gently into the grittier portions of technicality by introducing a flavour of background material and anecdotes; even the rough passages of pure theory and design are smoothed by straightforward explanations and diagrams. In other words, technical mystique has been reduced to common-sense description.

Following a chapter on the historical background of sailing ships, the first chapters describe the terminology and tools required in yacht design and lay down the basic foundations. Middle chapters go into the deeper waters of design such as defining hull shape, drawing the plans and performing the calculations. Later chapters discuss ancillary topics – finding the speed and power of a motor yacht, for instance, or how to design a rudder. There are also chapters that fill in any gaps of knowledge, such as a brief treatise on resistance and flow round hulls for those who have no clue of the subject. Useful tables and formulae are also included at the end of the book.

Because this is a DIY design manual, I have kept the physical and mental hardware required to an absolute minimum, while still catering for the more ambitious designer by including more sophisticated explanations where appropriate. In fact, for the more knowledgeable reader certain chapters may be skipped. But remember, advanced techniques or not, the contents of this book represent the hard core of yacht and boat design.

1
SETTING SAIL
An Introduction

A boatyard, unlike a car plant or factory, has a special atmosphere all of its own. The crazy scatter of timber and rope, half-built hulls and lumps of equipment and machinery, give it an appearance of a mad inventor's workshop. Yet its end products surpass Rolls-Royce cars in their sleekness and individuality. I started my marine training at one such yard, situated just off Portsmouth Harbour on the South Coast of England. It's here that the story begins.

When I was first shown round the yard by the Chief Designer, the place seemed to be a chaos of bustle and noise. Riveters' guns competed with boilermakers' hammers to deafen my ears; workmen crawled over the muddy stretches of the estuary, or high aloft on scaffolding. The scene was an awe-inspiring hive of activity that swamped my young senses.

'*That!*' said the Chief Designer, pointing his pipe at a dinosaur skeleton in steel, 'is a Girl Pat yacht. The frames have still to be plated.'

The machine-gun chatter of a nearby riveter shook any intelligent reply out of my head.

The next stop – the mould-loft – seemed a comparatively peaceful haven. Yet it left me bewildered as to why grown men should be tacking down long battens, gazing seriously along their lengths and then, after adjustments, gouging lines on their soft, plain black surface. These shipwright/loftsmen, I was informed, were highly skilled. Later, I was to learn the reason for their preoccupation with scribing lines on floors.

Next, I was conducted to the Naval Architect's office, where I saw my first yacht plan. The ink drawing on blue linen had artistic merit, in my naïve estimation.

'It looks very pretty, Sir,' I remarked.

The Chief grunted. 'It's a profile of the Girl Pat yacht you just saw. The hull is based on a trawler form. By the way,' he added dryly, 'a drawing is a product of design – not something to hang over your fireplace.'

I nodded humbly, though in later years he often had to rebuke me for putting unnecessary shading on cowl vents – or a flag on a Mast Plan. I still think that when you design your own yacht, it won't harm to include an artistic sketch or two.

In the main design office there were many bodies crouched over long drawing boards, all busy working on a variety of weird and wonderful plans. In order to reveal my keenness and intelligence to my future supervisor, I pointed to a sheet of paper covered with patterns of curved lines in three views, and asked whether it depicted the lines of planking on a hull.

'No,' he patiently replied. 'It's a Lines Plan – the body and soul of any yacht or craft.' He smiled. 'One day – with luck and training – *you'll* draw a Lines.'

It seemed remote at the time, but one day – after one or two efforts – I did. And many more afterwards. So will you.

That day I viewed a motley collection of other plans, confusing either in their purpose or in what they depicted. There were General Arrangements (GAs to the initiated), Construction Plans, Midship Sections, Bulkhead Drawings, Machinery Arrangements, and so on. One strange-looking beast on paper looked like a camel with tiger stripes. It was a Shell Expansion, I was told. This plan expands the hull in the vertical direction, so the deck-at-side line appears as a wavy line on paper. The 'stripes' were the longitudinal strakes of the hull plating. Its purpose was to show where the plates butted, what lengths were required, and so on. We need not bother with such a drawing for our designs.

For the next year or two I made tea for the whole office, ran off prints on an ancient carbon-lamp machine for the yard before developing them (and smarting eyes) in an ammonia box, and ran errands for everybody except the office cat. In between, I did some drawing work. As time progressed, so the menial tasks lessened and the drawing assignments became longer and more complicated. My theoretical knowledge grew by attending day and evening school, together with performing basic calculations such as tank calibrations and eventually, displacement and stability calculations. Five years on I emerged, trained in the fine art (and it *is* as much an art as a science) of yacht and ship design, one of the oldest crafts in human history.

To think that in bygone days the first naval architects were master shipwrights who worked with little theory, drawing their lines by eye and instinct.

Man's fascination with sea travel extends right back to Egyptian times and before. The propulsive mechanism in those days was muscle-power and oars. Then canvas and sail came on the scene and added a new dimension to the concept of travelling over water. The winds had been harnessed.

The Egyptians appear to have constructed proper ships as long as about 6,000 years ago, their hulls being fabricated from small, irregularly-shaped pieces of wood, due to that country's lack of large trees. The vessels were equipped with a square sail. The invention of the lateen sail in the Persian Gulf in AD 500 was a giant stride in sea transport because this triangular-shaped sail, set fore and aft, allowed for headway in head winds.

The Vikings also used a square sail and terrorised northern European waters, including the British Isles, as they struck at areas previously thought unreachable by sea in their longboats. One such longboat is illustrated in Fig. 1.1.

It was from the design of multitudes of smaller craft such as Hastings luggers and Brixham trawlers that the larger sailing ships evolved. Sailing

Setting Sail

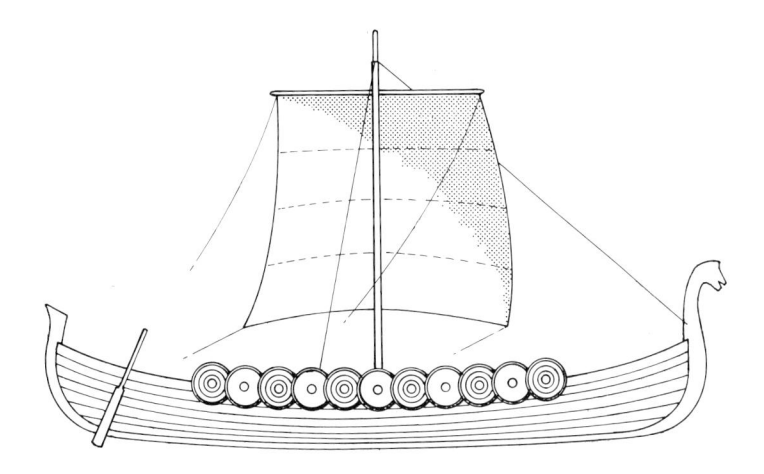

Fig. 1.1 Vikings also used a square sail and terrorised European and British waters in their longboats. One such longboat is illustrated above.

the seas in those days was not done for pleasure, but for purely commercial reasons – or war. The earliest picture of a three-masted vessel is the *Louis de Bourbon*, to be seen on the seal of Louis Bourbon (1466). The ship had a large mainmast, foremast, and a mizzen mast which carried a lateen sail, making it possible to sail in almost every type of wind.

Much is owed to the shipwrights of Spain, Portugal and Brittany for designing and perfecting such vessels, but the island race of Britain was also rapidly learning the fine skills of naval design – as much for survival and exploration as for profit. Shipwrights like the famous Phineas Pett were becoming conscious of 'shipshape' hull forms and when Pett designed the *Sovereign of the Seas* he curved its underwater hull to well up beyond the waterline. The vessel had four masts, her fourth carrying a sail called the 'Royals', probably because of her name.

Britain reached its height in sea power at the Battle of Trafalgar. Thereafter, its wooden ships-of-the-line ruled the waves for many years, but the days of wood and sail were doomed. In the wake of the Industrial Revolution they were superseded by iron and steam, then later, steel and diesel – even nuclear power.

With the decline of commercial sail (although it persisted in conjunction with steam power for a short time), the winds began to be used for pleasure and sport, but even before this, yachting as a leisure pastime had become somewhat established. It was in the European country of Holland that the modern pleasure yacht was first developed. The maze of waterways that criss-cross that small country were – according to pictures and illustrations – dotted in the 17th century with gaily carved and decorated shallow-draught sailing boats. The English kings James I and Charles I took a great interest in shipbuilding, while Charles II first learned sailing as a boy in a pinnace, and continued to indulge in the sport during his exile in Holland.

Plate 1 Wooden Ships-of-the-Line and sailing vessels ruled the waves for many years until they were superseded by iron and steel. The training ship *Sunbeam* shown above, is a fine example of the three-masted schooner.

The world's first sailing club, the Cork Water Club, was founded in 1720 and in the 19th century pleasure boating began to expand. Since then, yacht designs and hulls have changed considerably, as illustrated in Fig. 1.2. There were reckoned to be about 2,000 yachts in the British Isles at that time and 15,000 in the USA. The 19th century also saw the beginning of yacht racing as a great international sport.

In 1851 the schooner *America* crossed the Atlantic and took part in an international race organised by the Royal Yacht Squadron (RYS) for a 1000-guinea cup. Sailing against five British yachts, this Yankee greyhound won the race round the Isle of Wight by 19 minutes. The cup was taken back to New York and stayed there as a permanent ornament – until in 1984 an Australian yacht, with controversial fins, had the temerity to win the America's Cup, considered the most important yacht race in the whole world.

But there are also other exotic sailing competitions. For instance, in the summer of 1975 a fleet of modern ocean racing yachts sailed from the Thames at London, bound for Sydney, Australia, then returned again. Their aim was to beat the fastest times of the wool clippers.

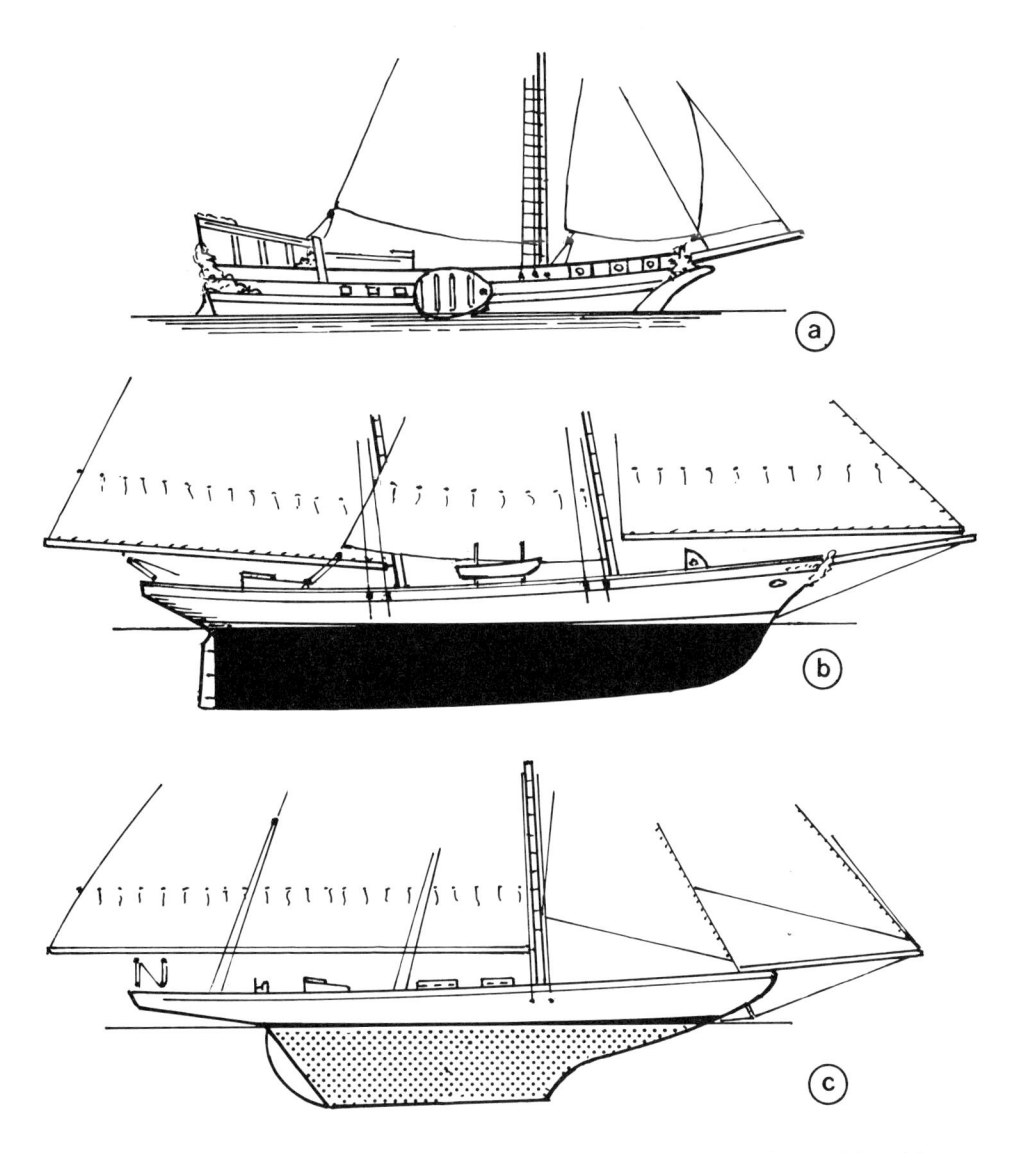

Fig. 1.2 Since the 17th century yacht hulls have changed considerably. Figure (a) is a 17th century shallow-draught Dutch yacht with leeboards, while (b) is a profile of the old US yacht *America*. Figure (c) is a J-Class racing yacht, one of the main participants in the America's Cup.

Yachting was the sport of the wealthy until recent times, when innovations such as mass production of fibreglass hulls (glass reinforced plastic, or GRP), together with a more affluent working and middle class, opened the seas and waterways to everybody who desired to feel a sea breeze on their cheeks.

Nowadays, sailing craft types are as numerous as varieties of baked beans,

Plate 2 During the transition from sail to steam, vessels were equipped with both forms of power, as shown above.

ranging from surfboards to trimarans, sailed for pleasure or competition and powered by canvas, motor, or both. Equipment and fittings have become highly sophisticated and technical, but the basis of design is the same as when Columbus set out to explore unknown lands, as later chapters will show.

2

SALTY TALK

An Explanation of Design Terms

Many jokes and pranks were played on trainees and apprentices in my day, the inevitable results producing equally predictable hoots of laughter from those in on the gag. They were usually well-tried and hoary old chestnuts that evolved around the jargon of the trade or profession.

For instance, I was once asked by a designer, wearing a suitably grave expression, to fetch him the 'key of the keelson' from the store – and while about it, to get him a bubble for his spirit level. The storeman roared with laughter at my request – and so did the whole office when I returned, red-faced and abashed. How was I to know that a keelson was part of a boat hull, let alone that the bubble was just *that*: a pocket of air in liquid. The instrument, by aligning the bubble in the middle, is used to find a true horizontal, liquid always lying in that position.

A lot of words or expressions used today have their origins in past events or practices. A number of these come from the old trades or crafts. Those who made their living from the sea – whether as sailors or builders – contributed much to the English language. For instance, the expression 'third-rate' (meaning 'of low standard') comes from the rating of the 74-gun wooden ships of Nelson's time, that being the most general rate employed in the Royal Navy. It is likely that the word 'waster' (a lazy and spendthrift type) originated from the poor and lowly '*waisters*' who formed the largest division of a ship's company and were considered stupid and 'without art or judgement'. These ignoble seafarers, as they were deemed, were berthed in the waist of the ship.

Before the 19th century the word 'yacht' usually referred to a publicly-owned vessel used by high-ranking personages such as kings, princes and ambassadors. Even in Egyptian times noblemen travelled up and down the Nile in state yachts; in fact, one yacht, a hundred feet long, was found buried in a tomb. Nowadays, a yacht usually means a privately-owned pleasure craft. Marine and yachting terminology has also been modified to some extent, due to more theory and development.

Yacht and boat design has its own technical jargon, much of it quite clear and comprehensible, but a number of terms are slightly bewildering. I recall during my initiation into the field being confronted with such suggestive words as *buttock lines*, *Body Plan*, and *fair curves*. At that precocious age it was a disappointment to discover they were only descriptions of part of the hull geometry – not the female anatomy. The *round of bilge* was *not* an order of cheap booze, but merely the curved connection between the side and bottom of a hull.

So that the reader will avoid such mistakes – and also because the understanding of certain terms is necessary to the design procedure – I have compiled a list of the major words and expressions used, together with their explanation. They are not presented in glossary form, but in readable fashion and as part of the natural process of yacht design. Only the more important terms are highlighted, others being given in the appropriate chapters, when necessary. They are presented in three sections: those relevant to the hull geometry, the construction and the calculations. Because a picture is worth a thousand typewriters (or words to that effect), I have relied heavily on the illustrations in explaining the terms.

Length is the most important dimension in yacht and boat design. For design purposes it is not the *overall length* (LOA) which the naval architect uses, but the length along the deepest level at which the vessel floats. This is called the *waterline length* (LWL), and it will later be seen why this is chosen. If a craft is to be built under the rules of Classification Societies such as Lloyd's Register of Shipping or other bodies, then the length used may be slightly different for large vessels, being measured still along the deepest waterline but terminating aft at some point such as a rudder stock or sternpost. This length is also called the *length between perpendiculars*, as will be explained later. The length used in Lloyd's Rules for yachts and small craft is defined in Chapter 23.

When a yacht designer talks about dimensions he is usually thinking of *moulded dimensions*. The moulded dimensions of a boat are those that are measured to the *inside* surface of the hull. The *maximum moulded* breadth (or beam) is measured at *amidships* and is the maximum width of the vessel *inside* the hull surface at that location. *Amidships* is located at exactly half the *waterline length* or *length between perpendiculars*. The *moulded depth* (again measured to the inside of skin) is the height at amidships from the very bottom of the hull (at the top of the keel) to the deck at side and, similarly, the *moulded draught* (draft) is the height to the deepest line at which the craft floats. The symbols and letters used for these terms and expressions are as follows:

LWL – Load waterline (deepest waterline at which the yacht floats)
LBP – Length between perpendiculars
⦶ – Amidships
B – Moulded breadth (beam), or the maximum width at ⦶
D – Moulded depth (to deck at side) at ⦶
d – Draught, the height to the LWL at ⦶

Soon we will be bringing all these terms together in two or three illustrations, but first we introduce some further words. The *sheer* of a yacht is the longitudinal curvature given to a deck when viewed from the side, while *camber* (round up) is transverse curvature and provides strength to decks as well as enabling sea water to be shed.

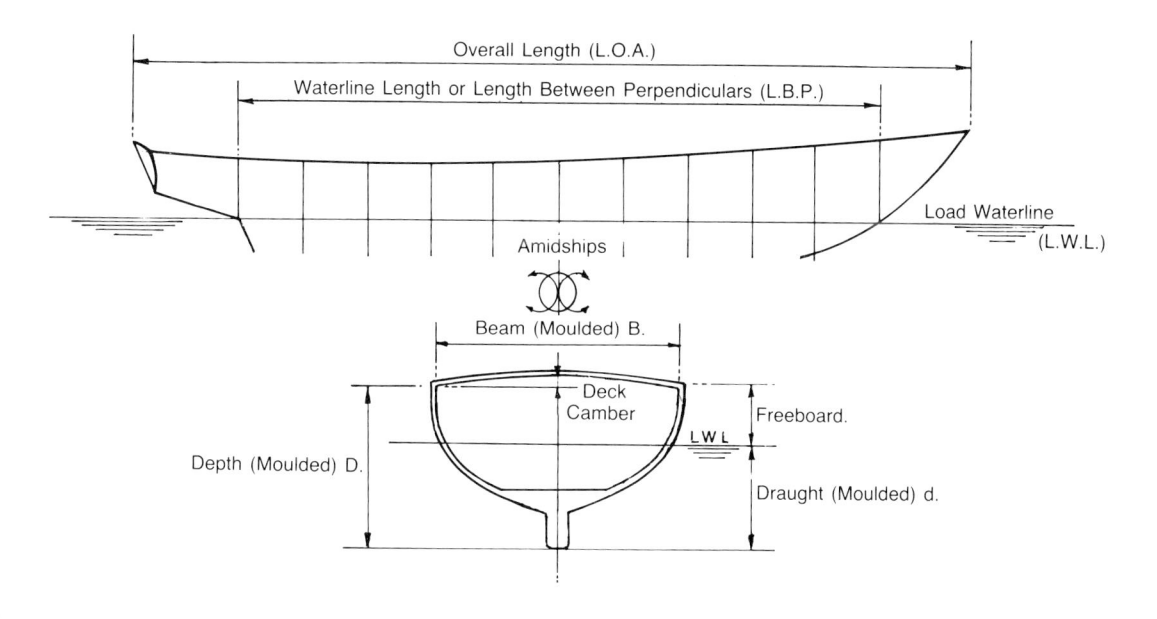

Fig. 2.1 Terms used to define the dimensions and other important parameters of yacht design.

Freeboard is the height from the LWL (deepest waterline) at amidships to the deck at side. All the above terms are illustrated in Fig. 2.1.

For a more detailed description of the hull geometry, imagine it to be divided by a three-dimensional grid. Viewing the hull in profile, it is divided equally by eleven vertical lines called *stations*. In certain cases the number of stations may vary to suit calculation and design requirements. The two end stations are located at the ends of the waterline length or LBP and are known as the fore and aft perpendiculars, respectively, hence the expression 'length between perpendiculars'. The number of ordinates (eleven for small craft such as yachts) is important to the calculations, as will be seen later. Now imagine the boat is sliced at each station and the transverse or athwartships view (when looking toward the bow) is drawn to an appropriate scale or size. The resultant plan is known as the *Body Plan*. As the sections at each station are symmetrical, only *half-sections* need be drawn and conventionally, the forward half-sections are drawn to the right, and the after ones to the left, both being on the same centreline.

The remaining two sets of lines in our three-dimensional grid divide the hull from stern to bow in vertical and horizontal planes. The horizontal lines run parallel to the waterline at which the yacht floats, those above being called *level lines*, those below, *waterlines*. When looking down on deck their sectional views form a family of curves starting with the deck curve and narrowing as they approach the keel. Once again, only a half side of the hull need be shown, as it is symmetrical in the longitudinal direction. Usually,

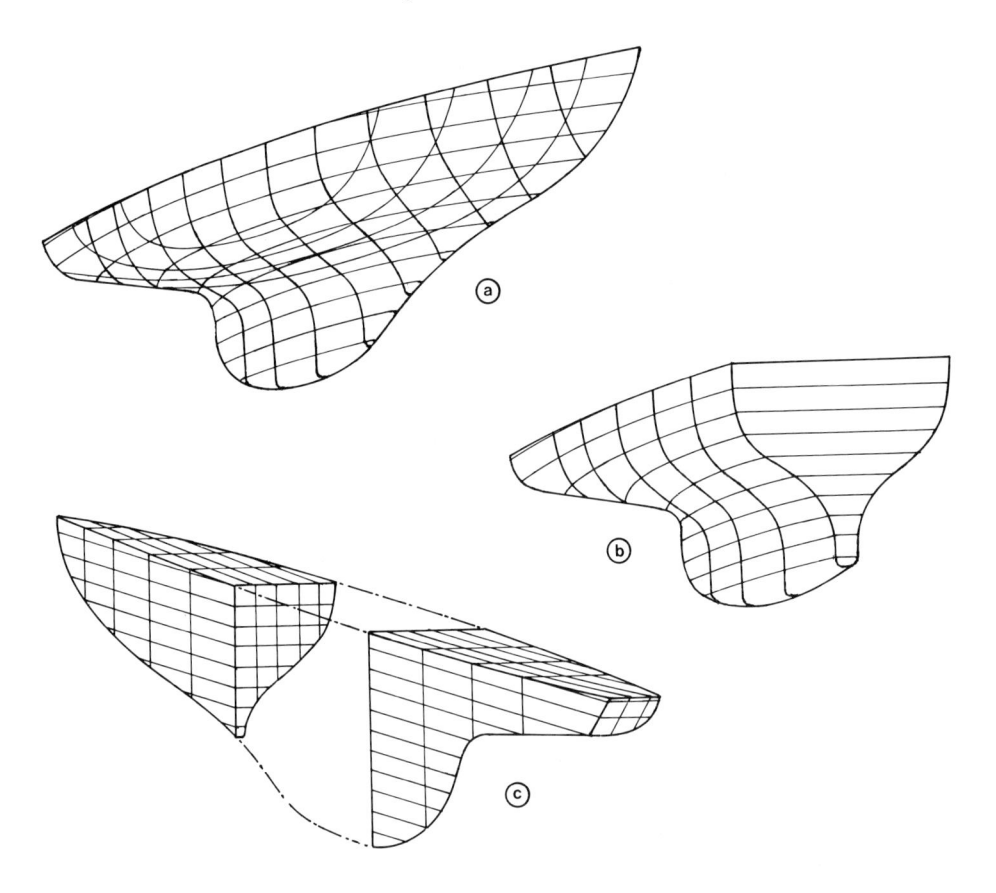

Fig. 2.2 A three-dimensional description of the geometry of a yacht hull. Figure (a) is a view of the whole and (b) shows the sections. Figure (c) illustrates the grid of bow, buttock and waterlines. Our DIY design will use such a grid to draw the Lines Plan.

the half-waterlines are shown above the longitudinal centreline, below being the half-level lines, and the whole forming the *Plan View*. For sailing yachts and hulls with great curvature, water lines and level are generally drawn on one side, and diagonals on the other. Figure 2.2 illustrates in three-dimensional fashion the various lines in which a yacht hull is divided. Figure 2.2(a) is a three-dimensional view of the whole; Fig. 2.2(b) shows the sections and Fig. 2.2(c) the grid for bow, buttock and waterlines.

The longitudinal lines which slice the hull in vertical planes form a set of curves that are profile sections of the hull from the centre outwards. The lines forward of amidships are called *bow lines* and those aft, *buttock lines*. The view is an *Elevation* of the hull which, together with the other two views, combines to form a drawing known as the *Lines Plan*.

There are a number of other words and expressions connected with the hull geometry. *Flare* is the outward curvature of the hull above the waterline, while *tumblehome* is the opposite, being the inward curvature.

Knuckle is a sudden change of curvature and *chine* is when the frame curvature changes abruptly at a knuckle. *Cut-up* is when a keel departs from a straight line at a knuckle. *Rake* is a small inclination to either the vertical or horizontal, and a *raking keel* is often given to small craft to ensure the propellers are above the line of keel, or to give sufficient depth to the sternpost. A *rockered keel* is one which has been given curvature so that it is deeper about amidships than at the ends, and applies to traditional long keel designs. A rockered keel is often the most efficient for small sailing craft. *Bilge diagonals* are lines on the Body Plan which aid the drawing process, while *bilge keels* are flat surfaces which project normal to the turn of bilge and reduce rolling. The *entrance* and *angle of entrance* are the waterline shape at the bow and the angle it makes to the longitudinal centreline. Finally, *trim* is the difference between the water draught forward and aft. Illustrations of some of these terms can be seen in Fig. 2.3.

We now come to the structural parts of a yacht or craft, beginning with *deadwood*, which is a vertical surface of the immersed hull. It has little width and provides maximum lateral resistance for little head resistance. Sailing vessels require deadwood to enable them to go to windward efficiently. The *keel* of a vessel is like the backbone of an animal and provides longitudinal strength. Other longitudinal strength members are *deck* and *bottom longitudinals, engine bearers*, etc. All the above are illustrated in Fig. 2.4.

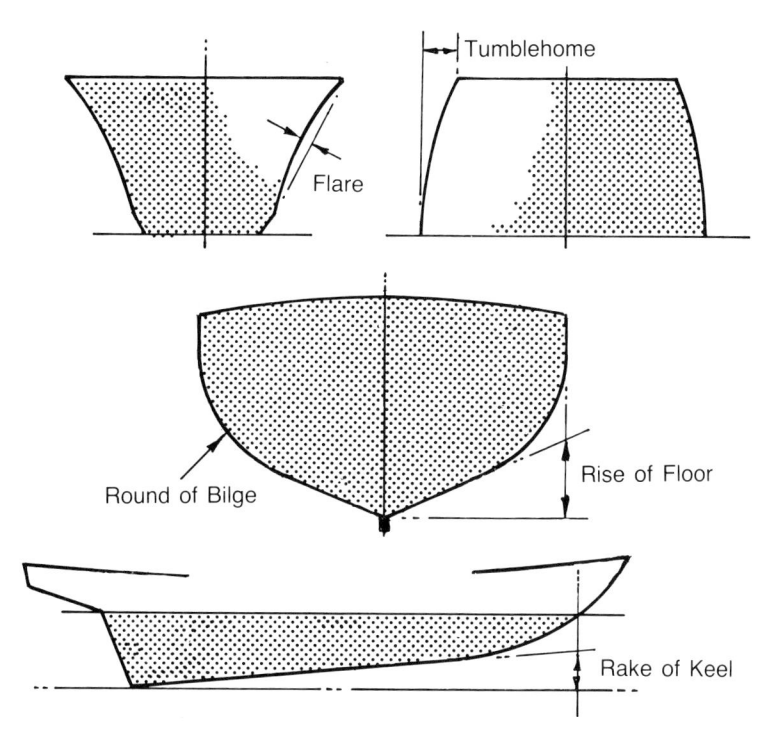

Fig. 2.3 Terms used to delineate the geometry of a yacht hull.

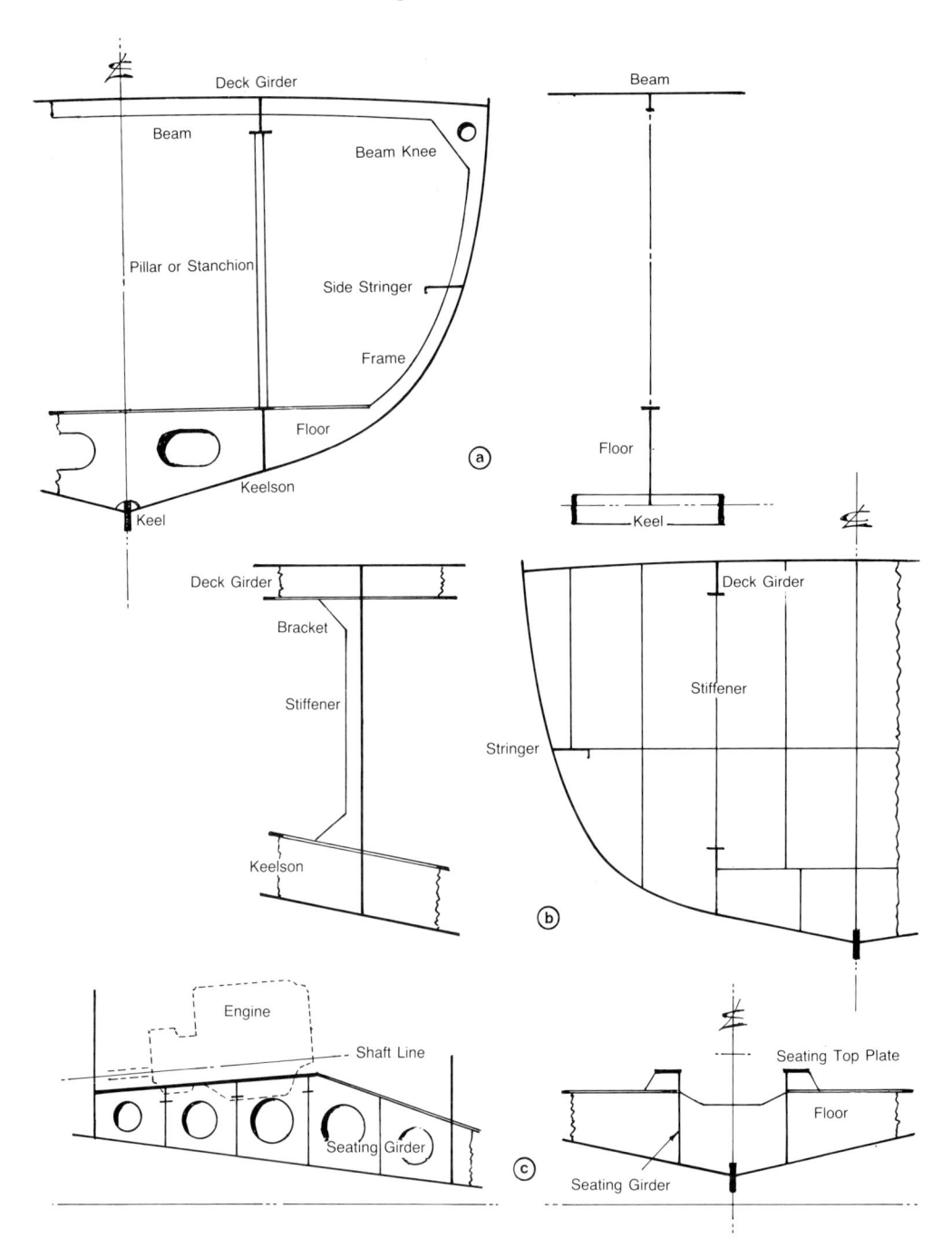

Fig. 2.4 Terms used to describe the constructional elements of a yacht hull. Figure (a) shows a typical section through the hull at a frame and (b) illustrates a bulkhead. Figure (c) is typical of inboard engine seatings or mountings.

Salty Talk

The transverse or athwartship members consist of *deck beams*, *frames*, and *floors*, with *bulkheads* to partition the various watertight and non-watertight compartments. These provide the transverse strength, again illustrated in Fig. 2.4.

The various connections between members are formed by triangular or shaped brackets such as *beam knees*, *floor* and *frame brackets* and *lugs*, while the skin or plating is stiffened by *stiffeners*. The complete framework forms a skeleton for the skin of the craft to be wrapped round and which, for wood craft, is either of *clinker* or *carvel* construction, but is nowadays more likely to be cold moulded or strip planted. Steel hulls are now welded and glass reinforced plastic (GRP) hulls are homogeneous.

Plate 3 The framework of a yacht hull forms a skeleton for the skin, consisting of frames, bulkheads, keel, longitudinals, etc. Cougar Marine's construction of the first of the two America's Cup yachts (above), is a perfect example of such a skeleton. (*Courtesy Cougar Marine*)

Finally, we come to the terms and expressions used in the design calculations which, unfortunately, introduce that horrible subject – mathematics. But be assured that all aspects connected with yacht design in this book will be of common-sense explanation and application.

Essential to all floating vessels, including yachts and small craft, is their *displacement*, which is the weight of water displaced by the vessel or yacht and given the symbol Δ. The *volume of displacement* (with the symbol now reversed, ∇ is used when defining the coefficients of hull form to follow.

There are a number of *coefficients* that are essential in determining the raw proportions of hull shape during the initial design stage. The most important is the *Block Coefficient* (C_b) and if you can imagine a block of wood having the dimensions of waterline length (L), moulded breadth (B), and draught (d), then the C_b is the ratio of the underwater volume of hull (∇) to this volume of block. In short:

$$C_b = \frac{\text{Volume of underwater hull}}{\text{Volume of circumscribing block}} = \frac{\nabla}{L \times B \times d}$$

The next comparison is the underwater hull to a block having a cross-sectional area of the underwater midship section. This is known as the *Prismatic Coefficient* (C_p) and is given by:

$$C_p = \frac{\text{Volume of underwater hull}}{\text{Waterline length} \times \text{Mid-section area}} = \frac{\nabla}{L \times A_m}$$

where A_m is the midship-section area.

The next coefficient involves the midship section area and is appropriately called the *Midship Section Coefficient* (C_m). It is defined by the ratio of this area to the cross-sectional area of the block used for the C_b. It is given by:

$$C_m = \frac{\text{Area of underwater mid-section}}{\text{Area of circumscribing rectangle}} = \frac{A_m}{B \times d}$$

Another coefficient of form is the *Waterplane Coefficient* (C_w), which is defined as the ratio of the waterplane area to that of the circumscribing rectangle and given by:

$$C_w = \frac{\text{Area of waterplane}}{\text{Area of circumscribing rectangle}} = \frac{A_w}{L \times B}$$

The final coefficient of form is the *Vertical Prismatic Coefficient* (C_{pv}), which is the ratio of the immersed volume to the area of the load waterplane multiplied by the mean draught, and given by:

$$c_b = \frac{\nabla}{L \times B \times d}$$

$$c_p = \frac{\nabla}{L \times A_m}$$

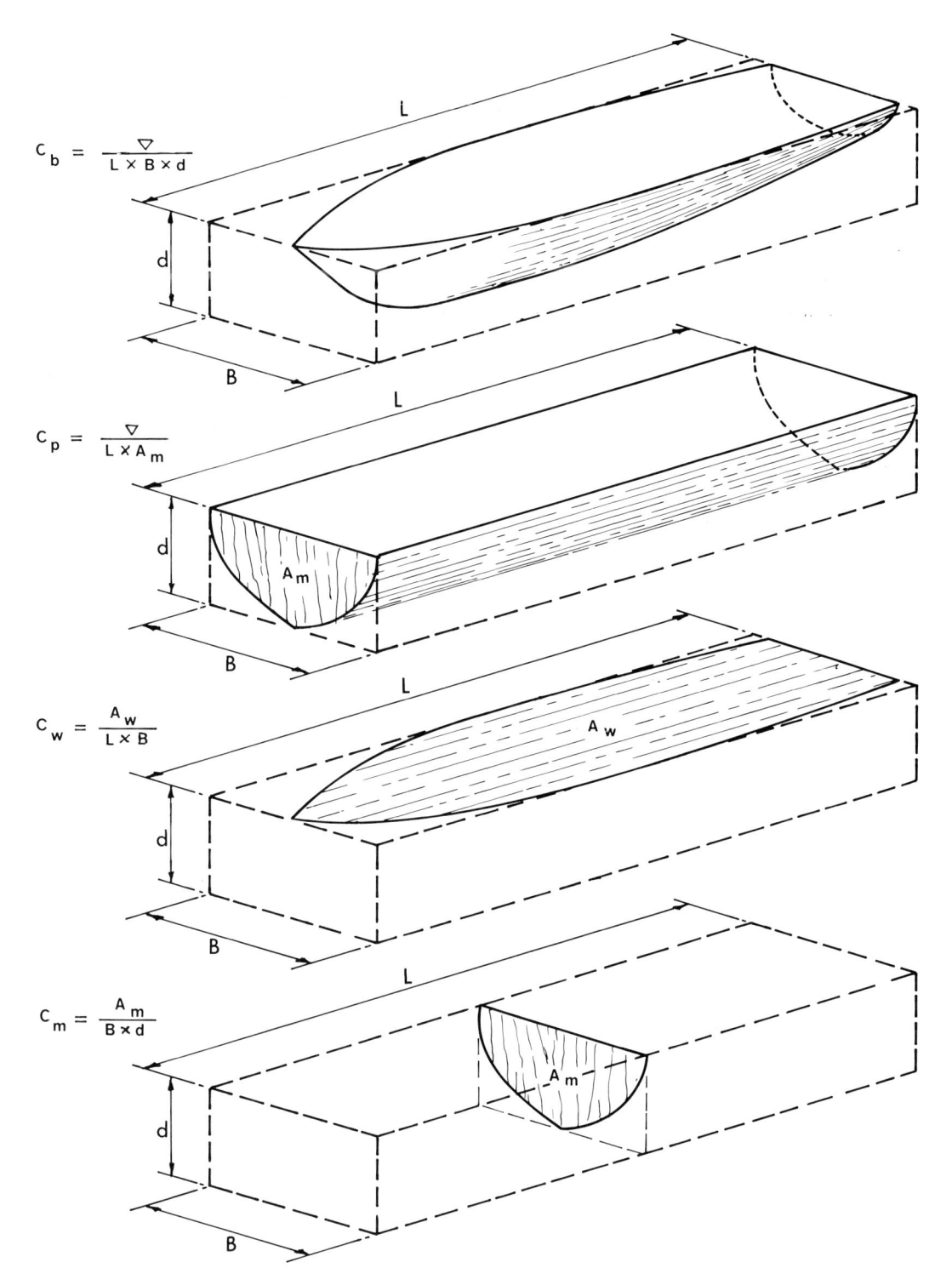

$$c_w = \frac{A_w}{L \times B}$$

$$c_m = \frac{A_m}{B \times d}$$

Fig. 2.5 The above illustrations show the principal coefficients of form used in the design of a yacht or boat hull. The coefficients are based on a circumscribing block round the underwater portion of hull.

$$C_{pv} = \frac{\text{Immersed volume}}{\text{Waterplane area} \times \text{Mean draught}} = \frac{\nabla}{A_w \times d_m}$$

where d_m is the mean draught.

C_{pv} is also given by:

$$C_{pv} = \frac{C_b}{C_w}$$

The prismatic coefficient C_p can also be simplified and related to the other coefficients to give:

$$C_p = \frac{C_b}{A_m}$$

All the above coefficients of form (except C_{pv}, which is rarely used in normal design) are illustrated in Fig 2.5.

A ratio that is integrally bound up with the speed and type of craft is the *Speed:Length Ratio*. This is given by:

$$\text{Speed:Length Ratio} = \frac{\text{Speed of craft}}{\text{Square root of length}} = \frac{V}{\sqrt{L}}$$

where V = Speed of craft in knots, and L = waterline length in feet.

Calculations for the above coefficients are shown in Chapter 14.

So we are now familiar with the more important terms and definitions of yacht design; it remains to show how they enter into the exercise. But first we need to talk about the tools of design, which is the subject of the next chapter.

3
BLOCK AND TACKLE OF DESIGN
Instruments for Drawing the Design

To apply a trade or profession requires some form of tools or instruments and the skills to use them. These need not be so complicated as one first imagines.

During my first week of training, the Senior Designer called me over to his board.

'You'll need to buy yourself some instruments, my lad,' he said, 'if you're going to do any drawing.'

'*Instruments!*' It sounded as if I were about to perform a critical operation or an important experiment.

'Let's see,' he continued, gravely. 'You'll require compasses, small and large; dividers, ditto; lines pens, thick, medium and thin. I suggest you also get a very fine one – to draw hairlines.'

I gulped. *Hairlines*!

'You will also need stencils – at least four sizes, including one large size for titles; 60° and 45° set squares, small and large; trammels, straight edge, French curves, sweeps . . .'

The list seemed endless, and to my discomfiture I noticed a small knot of other designers gathering round – grinning broadly.

'. . . a box scale, flat scale, slide rule, and rubber bands to keep your sleeves up when drawing. You won't have to buy a protractor or integrator; we have them in the office. Besides,' he added thoughtfully, 'they're rather expensive.'

'How – how much will the rest cost, Sir?'

'Oh, about £80 to £100.'

I couldn't even gulp. The sum represented two years' wages to me and would be equivalent to investing in a sophisticated business computer these days. I began to hate the profession I'd entered – especially when the others all went off, guffawing loudly. They had a merciless sense of humour in those days.

'Meanwhile, buy yourself two set squares and an adjustable compass,' said my kindly mentor. 'I'll lend you a scale. Now go and make the tea.'

The fact is, a bad workman usually blames his tools – but a good one mostly uses his head. You can get by with the minimum of instruments when designing your own yacht. Let's talk about the basic necessities first, then discuss more refined instruments where appropriate.

For drawing ship and yacht plans a fundamental requirement is to work on a flat, horizontal surface. This is because of the nature of the curves that have to be drawn. Professional naval architects and draughtsmen use long, flat drawing boards, but my first plans were executed on a drawing chest with just a sheet of cartridge paper tacked on top. For most yachts and motor cruisers, these can be schemed on quite a small surface – kitchen or dining-room table, for instance.

Horizontal and vertical straight lines come into most technical drawings and these are drawn using a tee-square and set squares. A tee-square, as its name implies, is an elongated T-shaped instrument made of wood, the small end butting against a side of board or table and the long edge used to draw horizontal lines. Set squares are triangular-shaped plastic instruments (90° at one corner and either 60°/30° or 45° on the sloping side) and when their 90° edges are butted against a tee-square, will enable the drawing of lines perpendicular to the horizontal lines (see Fig. 3.1).

A tee-square is very useful to the drawing exercise, but it does require a 'true' straight edge to butt up against. I remember how my own trainee self-esteem rose (at least, in *my* eyes) when one edge of the old plan chest was straightened by a joiner so I could use one. But bear in mind that it needs to be tight up against the edge to produce consistently parallel lines. I was rather careless at first, and ended up with efforts that see-sawed over the paper and made it seem as if the design was already on the rough seas.

Fortunately, most modern tables offer a good enough straight edge for our purposes, but if you do not wish to invest in a tee-square then two large set squares (a 60° and a 45° are recommended), butted on their sloping edges, will enable you to draw horizontal lines and perpendiculars to them. The technique is to draw an approximate horizontal, line up a side of one set square with this line (with the sloping side of the other butted to *its* sloping side), then slide up or down the guide set square to produce parallels or perpendiculars (see Fig. 3.2).

Marine design requires the drawing of a number of long, parallel straight lines such as waterlines, centrelines, etc. Set squares may be somewhat cumbersome for this task so a straight edge instrument would be of assistance. This can be any long strip of metal (aluminium alloy or light steel is ideal) about 2 inches or 60 mm wide, having one edge straight and true. Of course, one has to accurately measure equal distances from the ends of drawn lines, then lay the straight edge on the measured marks to produce parallels.

A yacht design also consists of many flat and sharp curves. Dealing with how to draw the former first, one common method is to use battens and weights. Battens are long slats of wood (or plastic, though I prefer wood) of varying lengths between 1 and 3 feet (30 to 90 cm for the metrically inclined), that are flexible and strong. Their cross-sections range from wafer thin to about ¼" by ¼" (about 6 mm by 6 mm) and are either constant or tapering. The tapered versions are used to draw those parts of waterlines that curve sharply at bow and stern. Weights are near/rectangular lumps of lead with a pointed wooden base that rests on the batten to hold it in place.

Block and Tackle of Design

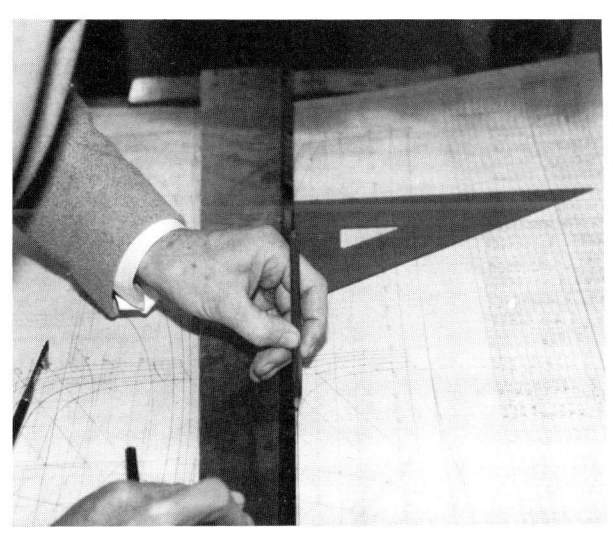

Fig. 3.1 A tee-square and set square can be used to draw horizontal and vertical lines for your yacht design.

Fig. 3.2 Another way to draw horizontal and vertical lines is to use two large set squares, butted together on their sloping sides.

Later on we will see how the shape of hull is determined, but for the moment let's assume we have a series of marked points on paper through which we want to draw a curve. The batten is pegged down by the weights as it is bent through – or as close as possible to – the points. This is shown in Fig. 3.3 and it should be noticed that the top of the weight rests clear of any interference when drawing the line. The surfaces of the batten should be smooth to allow for a similarly smooth travel of pencil or pen. Once pegged down, the designer will cast his eye along the batten to ensure that it runs 'fair', an expression used in design to describe a continuous curve without knuckles or bumps. Should there be a discontinuity, he will lift weights in the vicinity to allow the batten to spring naturally into a fair position (due to its elastic properties), then replace the weights in the new line.

It sounds quite easy, but in fact the technique of sighting the batten and laying weights in the right location takes a little time to master. I remember my first efforts resulted in the batten coiling away like a startled snake. Nevertheless, with practice one should be able to produce reasonable curves soon enough. I did!

Incidentally, does the process remind you of the antics of those shipwright/loftsmen mentioned in the introductory chapter? Yes, they were doing the same thing – full-size.

Having discussed drawing curves with battens – quite a sophisticated approach – it's fair to say that for yachts and similar craft we can draw most curves just as successfully with sweeps, French curves and pear shapes.

Design Your Own Yacht

Fig. 3.3 One way to draw yacht curves is to use battens and weights. The above picture shows the weights placed on the batten. They are then adjusted to rest clear and not interfere when drawing the curve.

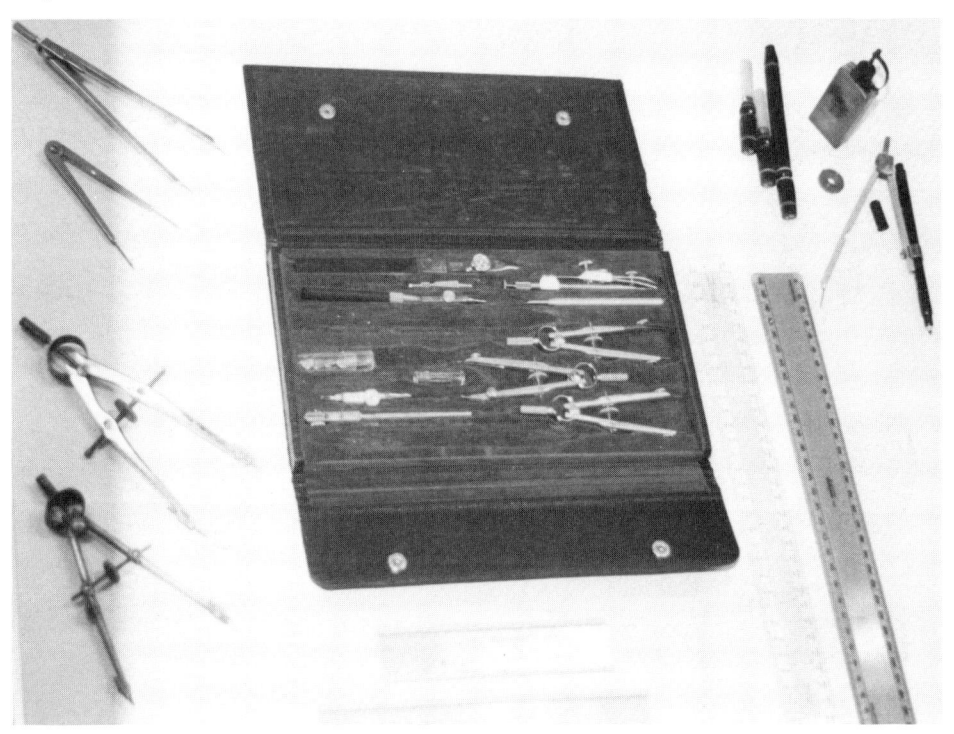

Fig. 3.4 A set of drawing instruments as shown in the picture is not really essential for our self-design purposes; a scale and compass (or circle templates) is sufficient.

Block and Tackle of Design

These plastic, geometric instruments can be bought in sets and are specially made to suit yacht and boat shapes, aerofoils and hydrofoils. They are easy to use: sweeps, because of their long, flat curvature, being suited to the mid-portions of waterlines in the Plan View, or else sheer line in the Elevation; French curves and pear shapes are ideal for half-sections in the Body Plan or the endings of waterlines at bow and stern. There are also a variety of other shapes such as ram's horns, cut-out ellipses, etc., as well as a proprietary device called the 'Flexicurve', which can be bent to any shape. The choice is yours, but only two or three sweeps and pear shapes – and possibly rams' horns – are necessary for our purpose. A selection of these instruments is shown in Figs. 3.4 and 3.5, while a planimeter is shown in Fig. 3.6.

Drawing instruments can be bought as a set (which is expensive), or

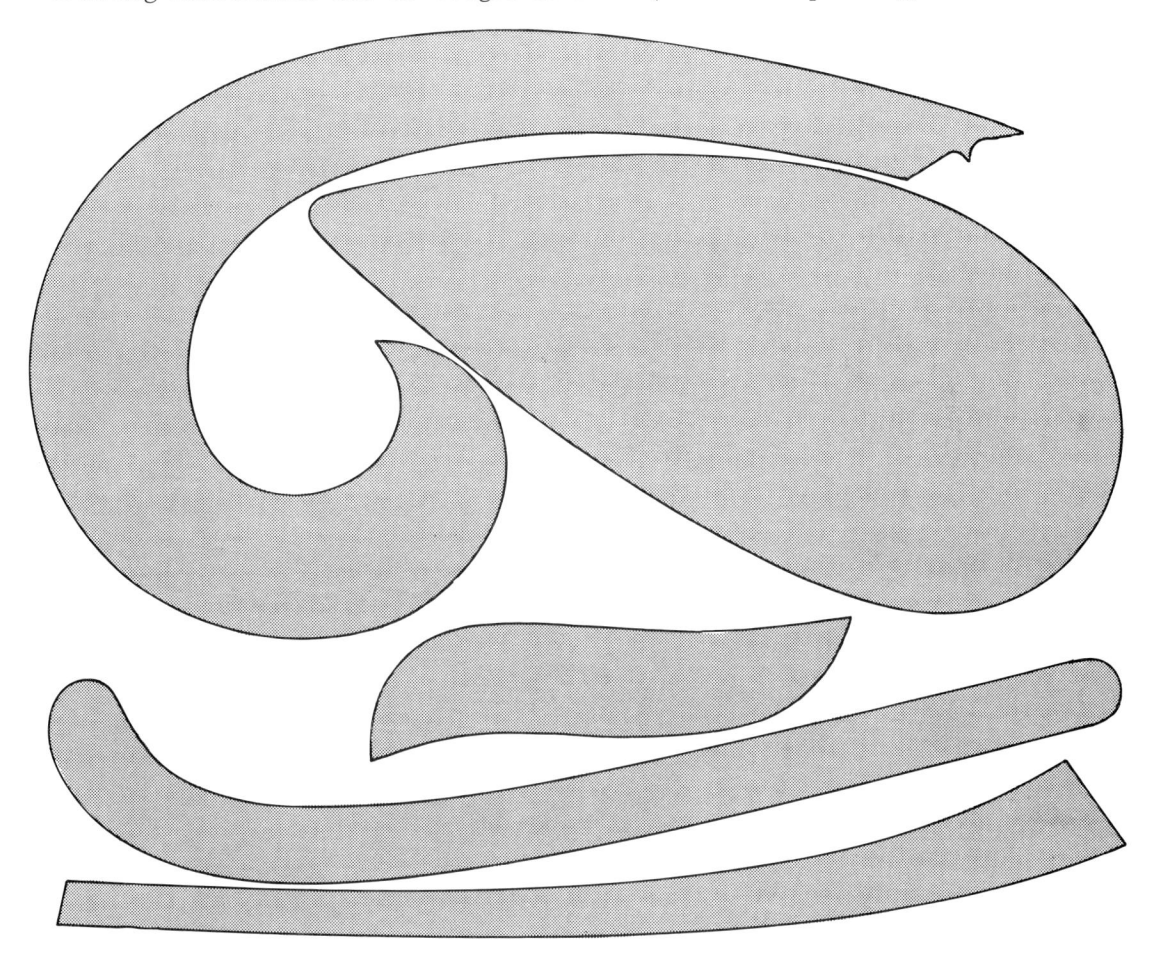

Fig. 3.5 Sweeps and curves such as those shown above are necessary in drawing the geometry of hull. A judicious selection of three or four types should be enough.

Fig. 3.6 A planimeter (as above) is a useful instrument for calculating yacht curves, but to avoid the expense, manual calculations can be carried out using Simpson's Rule, described in a later chapter.

individually, to suit requirements. The range of compasses in a set are used to draw ink and pencil circles which, for yacht design demands would be details such as portholes or holes in bulkheads and floors, hatch and doorway corners. Frankly, one pencil compass – maybe of the extension type – is sufficient to our requirements. When, late in my training, I bought a set of gleaming chrome instruments in a box of velvet padding, I belatedly discovered the expense was hardly worth the occasional twirl of a six-inch compass or the rarer event of tracing fine ink lines on blue linen (classy sets also have a range of lines pens for ink drawings). Finally, when using the compass, do not dig it into the paper, and use a light, twirling action when drawing circles and arcs. Templates for drawing circles are nowadays also available, to take the place of compasses. For calculation purposes a planimeter can be used, though this book uses a simple manual approach.

A scale ruler (just called a scale) is vital to all design work. There are two types: the flat scale and box scale. Both – as their name implies – have scaled divisions of feet or metres. For instance, an inch-to-the-foot scale (1 in 12) has inch divisions to represent feet, with appropriate sub-divisions for inches. A 1/10 scale for metres means every 10-cm division on the scale represents one metre full size (there are 100 cm in a metre), with sub-divisions providing the centimetres and millimetres.

A flat scale (which looks somewhat like a school ruler) usually has four scales, two either side. Some introduce eight by using the ends. A box scale has a triangular cross-section incorporating at least six scales. I first opted for a flat scale, then felt quite cheated when I saw a box scale with its extra offerings, but the fact is, a flat scale is easier to use and covers most of yacht design requirements.

So what range of scales are suitable for our task? Yachts between 10 and 20 feet (about 3 to 6 metres) are comfortably handled by from 1" to 3" to the foot ($1/12$ to $1/4$) scales, or $1/10$ to $1/5$ for metric dimensions, depending on type of drawing and space available. Large plans such as General Arrangements, Lines Plans, etc., need to be drawn to the maximum scale allowable, but will obviously be at the lower end of the range; detail drawings are at the upper end. Using the scales is very simple and needs no explanation, but be careful that you measure and mark accurately, making just small dots or ticks on your paper.

We now come to the pencils required for the job. These are graded by letters in relation to their hardness or softness of lead, the two most commonly used for design being in the H or B categories. A 2H pencil is most suited for drawing lines and an H or HB for printing instructions. Sharpen your points to a 'chisel' edge rather than the usual 'dagger' point. This enables cleaner and finer lines to be drawn. Sharpening pads can be bought for this purpose; a piece of fine sandpaper or the striking side of a matchbox will also serve.

Equipping oneself with all the items mentioned above might appear to be slightly expensive. But the total investment should be no more than about £50 ($70), and considerably less if purchasing cheaper instruments. And remember, once bought, they can be used for many years – and for many designs. If even *this* investment is pocket-shattering, then probably the only essential is a scale and lots of ingenuity in devising makeshift aids to drawing straight and curved lines.

So we arrive at the stage of putting pencil to paper. But wait! What *type* of paper? For the major plans good quality tracing paper is recommended, not too thin or it is likely to tear during the drawing process. The Lines Plan has to be drawn on something more substantial such as cartridge or thick-backed paper. This is because the plan has to remain as stable as possible, free from shrinkage or expansion. Later, we will learn that a successful design partly depends on its accuracy. For copying and re-drawing purposes it can always be transferred onto tracing paper, along with a table of accurate dimensions called *offsets*. A more expensive, but more stable and durable material is plastic film, from which prints can be immediately taken.

Many lesser drawings can be just sketches, drawn on inferior materials, while final plans such as the As-Fitted General Arrangement are usually preserved as a permanent record in Indian ink on special quality plastic film (blue linen in the old days). Though such drawings are not essential, some readers may become design-proud enough to lavish much time on their end product.

The paper, cut to a size that allows for the scaled views to be represented, is pinned or tacked down onto the board or table (masking tape is very suitable) in a secure manner. The procedure is to anchor the top right-hand corner, sweep a hand along the paper to the bottom left, then secure this corner. The same is done with the two remaining corners to leave the paper lying as tight as a drum skin – *but we don't start to draw immediately.*

Let the paper stand for an hour or two (even longer for Lines Plans) to allow for any shrinkage or expansion caused by temperature and humidity. Afterwards, take in any slack there is, re-fastening it to its more stable position. Of course, with film this is not necessary. We are now ready to start drawing.

Not quite!

One minor – but absolutely essential – drawing aid is an eraser. Make sure you buy a soft one for your pencil drawings. I once worked on a plan which took so long to execute that it began to look like a cabbage patch. In a fit of tidiness I began cleaning it up with a hard green rubber normally used for ink drawings. It ended up a streaky mess of grey and white – a ghostly Zebra of a yacht hull.

So now we are ready to begin. Many a potential writer has looked at the blank sheet of paper in his or her typewriter with sinking heart. Something intelligent, interesting – even informative – has to appear on that monstrous little plain of white. As you look down on your neatly secured sheet of drawing paper you may also suffer such agonies (*I* did – even unto the fifth generation of plans).

But not to fear! Designing your own yacht is *not* difficult – provided you are clear in your aims. Such aims may have to be revised or jiggled around during the course of design, but a logical formulation of your thoughts is of immense assistance to the process – and needs to be written down on paper. This is called the *specification* – a log of your design thoughts, if you like – and is the written guide from which all else follows. It is also the next step toward the creation of your own design. But before we write this design log, it will help to know what we are going to list and itemise – which the next chapters will review.

4

CHARIOTS OF SEA AND SAIL

Types of Yachts

Before going further into our DIY design, it will be worthwhile to survey the various yacht types and their service.

The evolution of marine transport in early days was a slow and fragmented process. The wooden chariots of the sea developed over aeons of time, finally culminating in the four-masted galleons with billowing sails. Then this century saw an explosive advance in all forms of shipping and pleasure boating, particularly after the Second World War. New innovations, materials and designs have completely changed the marine scene, throwing up some floating oddities in consequence.

Plate 4 Many strange marine beasts have been designed by naval architects, such as the marine meteorological observation ship, *Chofu Maru*, above, with its variety of latticed stacks on deck and superstructure. (*Courtesy I.H.I.*)

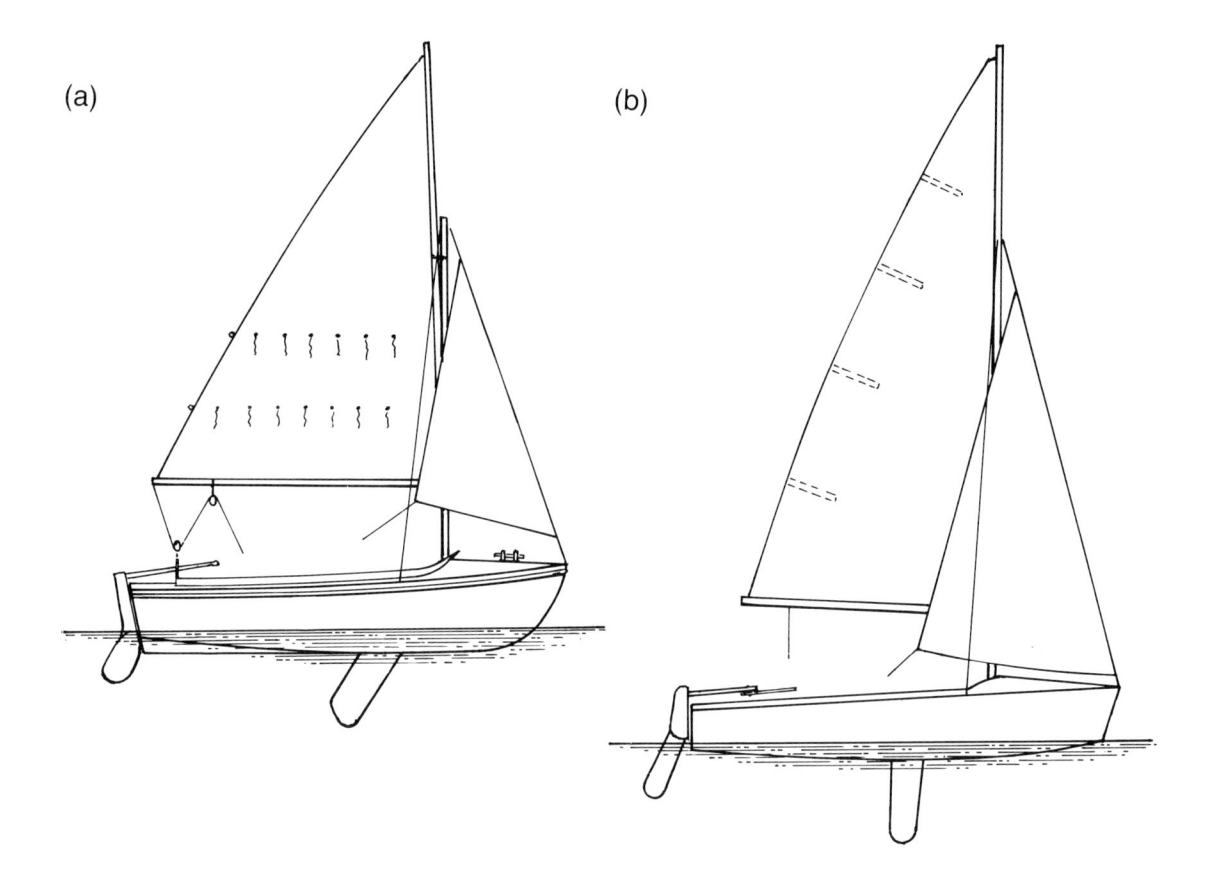

Fig 4.1 The two most common and popular types of dinghy are (a) the general purpose dinghy, being suitable for the novice, comfortable and easy to handle, and (b) the racing dinghy, requiring skilled yachtsmanship, muscle-power and not so much regard for comfort.

There are four principal types of boat in the sailing yacht stable, the most popular – and inexpensive – being the centreboard dinghy. Dinghy sailing can be done for sheer pleasure, or racing, usually on lake, river, or else offshore. The general-purpose dinghy is comfortable and easy to handle, being equipped with a small sail. Speed is of little consequence for this type, which is eminently suitable for novices. Some day-boats may be included in the racing type, which ranges from the 7'-6" Optimist to the 20' Olympic Flying Dutchman. The pure class racing dinghy is only suitable for skilled yachtsmen, requiring muscle-power, and lacking any refinements of comfort. Nevertheless, with its large sails it can achieve very fast speeds. The two types of dinghy are illustrated in Fig. 4.1.

Next in the stable is the cruising yacht, which needs to suit the area it will sail in. Design considerations for such a yacht are strength and accommodation, and one good idea is to have the deck as strong as the planking to resist

Fig 4.2 A number of design considerations enter into cruising yacht design, such as strength and accommodation. The 26'.0" Newbridge-built family cruising yacht, *Pioneer*, above, has six berths and an inboard diesel engine. (*Courtesy Newbridge Boats Ltd.*)

the heavy seas it may encounter. Small cruising yachts will have sitting headroom, as against the larger types which have full headroom. Modern cruising yachts have a large beam to allow more generous accommodation below decks. Another good design idea is to incorporate a self-draining cockpit to avoid pumping out when shipping water. If long sea voyages are envisaged, room should be allowed for in the design to take a dinghy on board. A typical cruising yacht is illustrated in Fig. 4.2.

In all yachtsmen's eyes a step up in class is the racing yacht, which should, even more than others, suit its area of service – otherwise there may not be appropriate competition for it. The primary choice of keels for such craft is centreboard or keel, the former being suited to smaller, shorter vessels. Ocean racing is the most glamorous form of competitive yachting. It is an expensive sport and for those readers who are prepared to put their marine fantasies onto paper, be prepared to look very carefully at the rules and handicapping requirements involved, as well as designing a large enough vessel to carry crew, equipment and stores for many days; in truth, it is a small sailing ship. Figure 4.3 shows a fast, large, ocean-going yacht.

Design Your Own Yacht

Fig. 4.3 If you want to put your design fantasies on paper then ocean racing is a glamorous, but competitive form of yachting. One famous racing type is the 12-metre yacht above, built by Cougar Marine. (*Courtesy Cougar Marine.*)

Fig. 4.4 There are a variety of motor yachts that ply the seas and inland waters. Some are yachts with auxiliary engines and others are purely engine-powered. A selection is shown above.

Plate 5 In the pleasure boat field there are sedate river and estuary cruisers. Many are used for floating holidays, such as the *Connoisseur* Cruiser above, employed by Hosseasons for UK and continental waterway vacations.

Motor yachts and cruisers are built mainly for pleasure, although there are racing designs such as the half-decker that planes its way round harbour or channel. The offshore racing or powerboat may have some form of accommodation to comply with rules, but the priority is power and speed. In the pleasure boat field there are sedate river and estuary cruisers (some no more than a form of mobile accommodation along these waterways), offshore motor yachts with streamlined superstructures and gleaming chrome, and even more luxurious ocean-going yachts that are virtually mini-palaces of the sea. There are also serviceable craft that are designed for a purpose, such as fishing or other marine leisure activities. A variety of motor yachts and cruisers are illustrated in Fig. 4.4.

We can now decide whether our design is to be a simple dinghy – or a millionaire's floating palace. Whatever we choose, the design has to be rigged, powered and equipped, and the following chapters aid us in this task when writing the specification.

5

TIMBERS THAT KISS THE WATERS
Materials of Yacht Design

Some strange materials have been used for designing vessels in maritime history, including bunches of reeds to make the first rafts, balsa from the upper reaches of the Amazon and papyrus, growing in abundance some six thousand years ago on the banks of the Nile. In recent times iron, steel and even concrete have been used to construct merchant ships and smaller craft, while plastic materials such as GRP have completely changed the yacht construction scene.

Much of the oak used in building the ships-of-the-line in Nelson's time was grown in the royal forests such as the Forest of Dean and the New Forest. The best wood was not considered suitable for use until a year after it was cut. The keels were made of elm, the frames of oak, and the planking and much of the other timbers, of teak. These were the fine timbers that kissed those waters in historical times. Nevertheless, due to exposure and constant wettings and warpings, many ships rotted to pieces after only a few months at sea. In fact, the general life of a ship in those days – under such conditions – was only eight or nine years. These days, with better materials, treatment and maintenance, hulls last well beyond the life of their owners.

GRP's entrance into the yachting scene repelled many traditional yachtsmen, whose inherited instincts for wood made any synthetic material a sacrilegious intrusion into the purity of yachting. Yet it soon became obvious to yacht designers and builders that these new plastic materials opened the door to cheap mass production of standard hulls, combined with other advantages such as lighter weight for the same strength, minimal building costs and ease of maintenance. Little wonder then, that GRP has become the ubiquitous material in yacht and small boat construction.

Other advantages of plastic materials for marine purposes are that they are, in general, in good supply, not subject to galvanic corrosion, dry rot or the attacks of marine borers such as *teredo* and *grabble*. They also do not require a highly skilled labour force in their use and construction.

The two divisions of plastics commonly used in industry are *thermoplastics* and *thermosets*, the latter being the only one suitable for marine purposes. Both require heat application during manufacture, but with the thermoset type the change from plasticity to rigidity is permanent and cannot be reversed by reheating. Polyester resins require only contact or low pressure, and no external heating is needed as this is supplied by exothermic reaction from the chemicals used to complete the 'curing'. With the addition of glass

Timbers that Kiss the Waters

Plate 6 Glass reinforced plastic (GRP) has become the ubiquitous material of yacht and boat building, being of homogeneous construction and suitable for mass production. Above is a shot of Cougar Marine's GRP facility. (*Courtesy Cougar Marine.*)

fibre cloth or mat as reinforcement (polyester resins in themselves have no great strength), the bonded material becomes very suitable as a marine material of construction.

Another modern material for yachts and small craft is aluminium alloy. Pure aluminium is quite resistant to corrosion but, unfortunately, is too soft for constructional purposes. Alloys for marine use combine good resistance against corrosion with the greatest compatible strength. In an alloy-steel connection the alloy will corrode first, being lower down the electrochemical scale, so special precautions are required when such contact occurs. When comparing the strength:weight ratio of alloy to steel, a rule-of-thumb guide is that an alloy hull will be 50 per cent lighter than its steel counterpart for the same strength – a considerable advantage. On the other hand, it is a much more expensive material than steel or GRP.

Design Your Own Yacht

The first iron ship built was the *Auron Mumby* (1822); she was also the first prefabricated construction, being put together on the Thames. Chemists then discovered that adding a small amount of carbon to iron turned it into a tough and durable metal – mild steel – far superior to other materials in marine construction and in other applications. To this day mild steel is pre-eminent on the shipbuilding scene, as well as for certain yachts and small craft.

Lloyd's Register of Shipping and other Classification Societies have tight rules regarding the grading of mild steel for marine construction; Lloyd's have five grades designated A, B, C, D and E. Mild steel is a heavy material compared to those previously mentioned, but it is relatively cheap and easy to fabricate, especially with the modern plant machinery now available. Nevertheless, it has to be protected against corrosion and marine growth by using the right treatments and paints, and probably requires more maintenance than GRP or alloy.

If your design dreams are of a vintage nature then wood has an appeal built on centuries of craftsmanship and tradition – not forgetting nostalgia. A wooden yacht hull is also a thing of beauty in its varnished state. The two types of construction when planking a hull are carvel and clinker-built (see Fig. 5.1), the former offering a smoother surface but requiring more skilled workmanship. Of course, cold moulded and strip planking are more modern methods of wood construction.

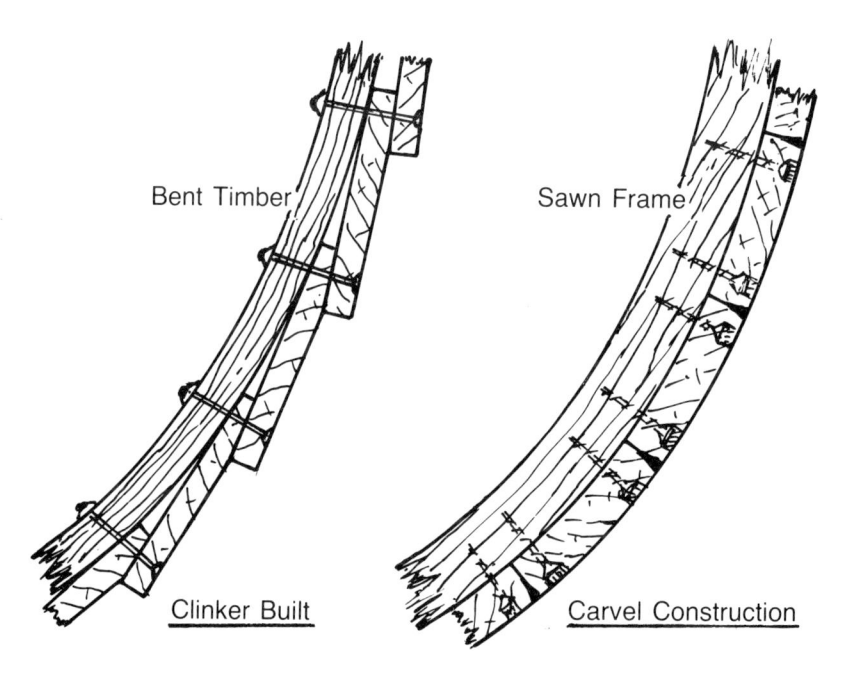

Fig. 5.1 The two types of construction when planking a hull are carvel and clinker. More modern methods are cold moulded or strip planked.

Timbers that Kiss the Waters

The desirable qualities of wood in marine and yacht construction depend on the use to which the wood is to be put. The bent frames in a small craft require a wood which can be easily bent when steamed. Deck planks should be resistant to impact and abrasion and wear well, while keelsons should be tough, strong, and able to take end fastenings. Side planking (especially near the waterline) should be able to resist continuous wetting and drying, while wood in covered conditions should resist decay due to a poorly ventilated environment. In all cases toughness and resilience are desirable, (but not brittleness), together with minimal shrinkage or swelling. Unfortunately, no timber can fulfil all these conditions.

Teak, oak, and pitch pine are the most suitable for general yacht – and boatbuilding purposes. Teak is extremely durable and resistant under all conditions and has a natural oil which preserves it against both weather and fungal attack. It is one of only two woods which can be left unpainted or unvarnished without harm. Oak is ideal for sawn timbers, knees, keel, etc., while pitch pine is the best of the conifers for yachtbuilding purposes, being

Plate 7 Oak is ideal for sawn frames and other constructional members of sailing boats, as exemplified in the traditional form of construction for the above fishing vessel.

Plate 8 A clinker-built boat built by a specialist craftsman in his own back garden. Carvel construction requires even more skills.

only inferior to teak for side planking and other hull parts. There are other woods – and variations – many not suitable to yachtbuilding purposes, as well as wood substitutes, but it is not possible to discuss these in the present book.

Finally, there are plywoods and laminated constructions. Such materials offer a cheap method of construction for small yachts such as dinghies and simple river cruiser designs. Bulkheads also form a very useful application of plywood as does, in some cases, a deck. Plywood frames are wasteful of material and laminated or built-up frames are to be preferred in craft just too big for all bent frames. In laminated construction the plies are not laid diagonally or in opposite directions, but the grain runs the same way in different layers. Special plywoods under various trade names are compressed in manufacture, and for all plywoods used in marine construction it is essential they should be bonded with one of the synthetic glues.

Should the timbers of your design actually kiss the seas, lakes and rivers, you will have added a small knot to the long, long road of yachting and maritime history – and the pride that goes with it.

6
UNDER CANVAS
Sails and Rigging

Sailing craft, large or small have a singular form of beauty. Given some analysis, it will be seen that their picturesque quality is concentrated on the sails that curve so graciously under the wind's breath.

The history of sail goes back thousands of years and covers world-wide territory. The very earliest sails – before weaving was invented – appear to have been made of leather, stitched hides being shaped into a square sail. The Egyptians used flax for sail material in about 3000 BC. It was then introduced into Northern Europe at around 500 BC by the Romans and came into general use up to the end of the sailing era. Early Viking ships appear to have used woollen sails and Egyptian cotton was used for yacht sails, while American cotton was commonly employed by American square riggers and schooners. Now, with the exception of a few historically preserved ships, all sail cloth is synthetic – mostly Terylene or Dacron.

Early sailing vessels were only capable of driving downwind; then the Lateen Rig enabled them to be more versatile in adverse winds. In Atlantic and North European waters the square sail developed from its primitive form to combine with other sails until the Ship's Rig of three masts with two square sails, a lateen mizzen and later, a square sail under the bowsprit, became standard until the end of the era of sail.

Having executed professional artwork on marine and yachting subjects, I find the most fascinating element in a yacht scene are the sails. A yacht under canvas lends itself beautifully to the sweep of an oil brush or pen. Obviously, a sailing rig is also very much part of the design process.

Before discussing the yacht rig types let us briefly go through the various parts that compose a rig. A rigging can be divided into *standing* or *running* rigging. Standing rigging helps maintain the mast in a permanent position and stands throughout the season. The *stays* provide fore-and-aft support for the mast, while *shrouds* provide lateral support. Small boats may have just two shrouds, leading slightly aft of the mast, and a forestay. Taller masts have more complex standing rigging.

Running rigging controls the movement of masts and spars and is continuously being hauled or slackened through blocks and tackle. Running rigging consists of *halyards*, which hoist the sails while under way (usually rope for smaller craft and flexible wire for larger ones); *sheets*, which control the sail angle (usually of rope); and *topping lifts*, which take the weight of the main boom when hoisting, reefing or lowering the mainsail. There may also be other rigging lines for more complex rigs.

Design Your Own Yacht

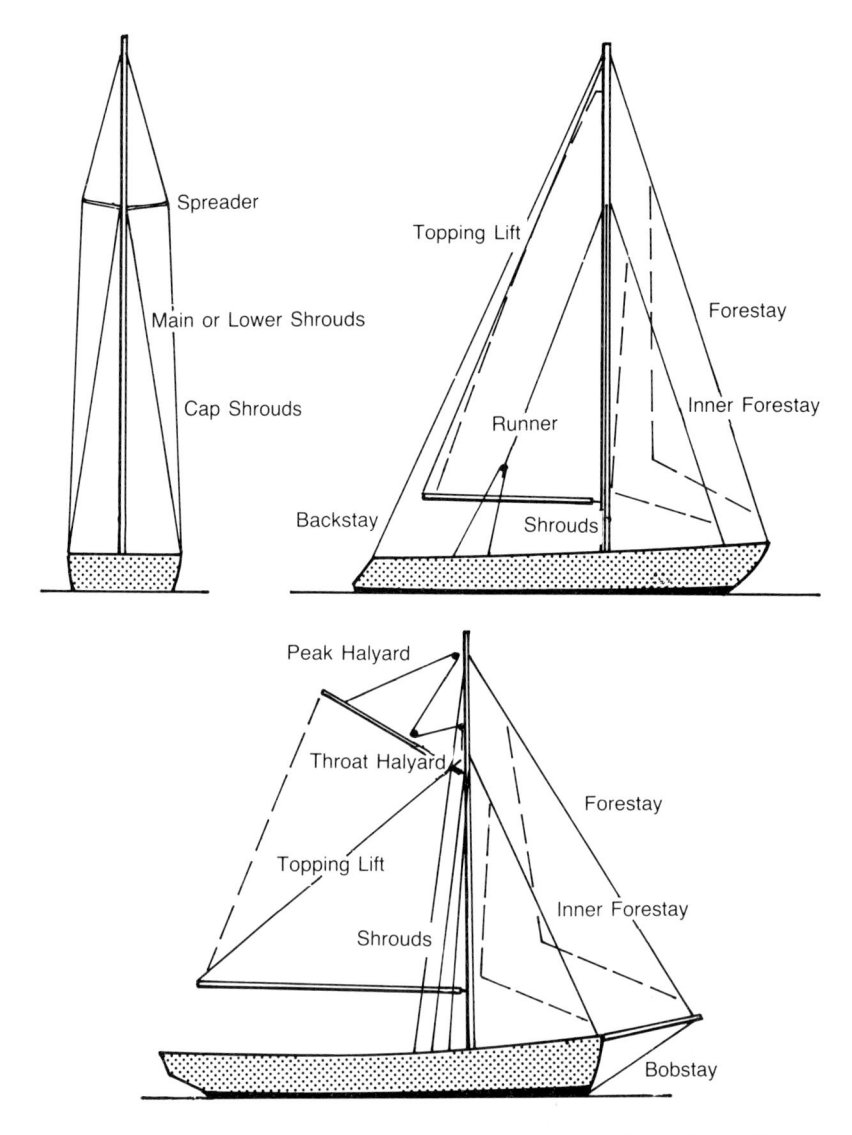

Fig. 6.1 There are a number of elements to the rigging of a sailing vessel or yacht. A scheme showing the important ones is illustrated above.

Masts may be of metal (usually aluminium alloy these days) or wood. Large yachts may have two – possibly three – masts, but yachts for the average enthusiast have one mast. There is a spar called the *boom* attached to the mast, pivoted to move freely. On gaff-rigged yachts will be a gaff, also of pivoted attachment. Yachts carrying a *spinnaker* will carry a *spinnaker boom*. Some vessels carry a spar called a *bowsprit*, which projects from the bow to attach the top mainstay (sometimes called the *forestay*). A scheme showing some of the above rigging elements is illustrated in Fig. 6.1.

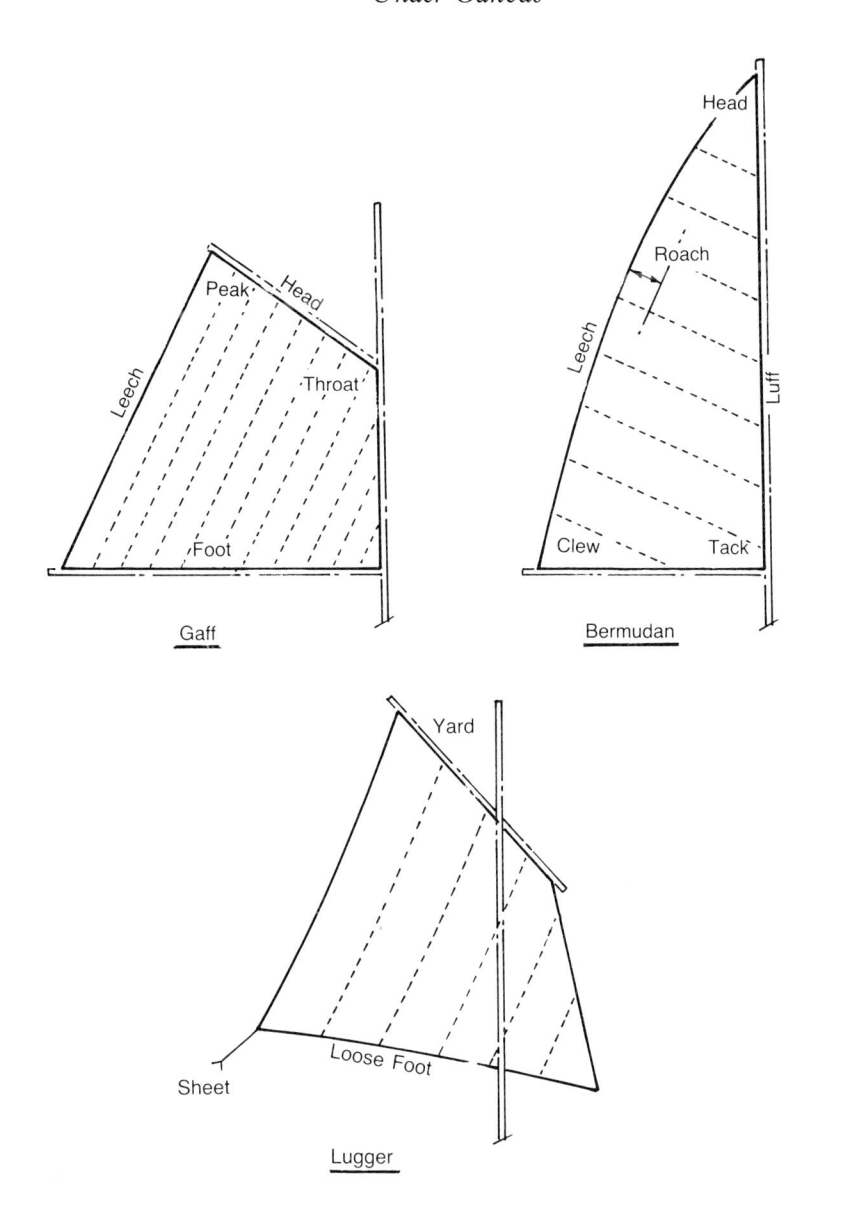

Fig. 6.2 Two major shapes for yachts are the Bermudan and gaff, while dinghies may have a lug-sail.

The two major sail shapes for yachts are the *Bermudan* sail and *gaff* sail. These are shown in Fig. 6.2, together with relevant names of their corners and sides. The Bermudan (triangular) sail gets its name from the 'leg o' mutton' shape of earlier vessels indigenous to British waters. In the gaff rigger, the gaff spar at the head extends the top edge of this fore-and-aft sail to give it its special shape and name. The working sails for boats going to windward in moderate to fresh breezes are a *mainsail* and *headsail* for sloops,

Gaff Cutter

Gaff Yawl

Gaff Ketch

Bermudan Sloop

Bermudan Cutter

Bermudan Yawl

Bermudan Ketch

Bermudan Schooner

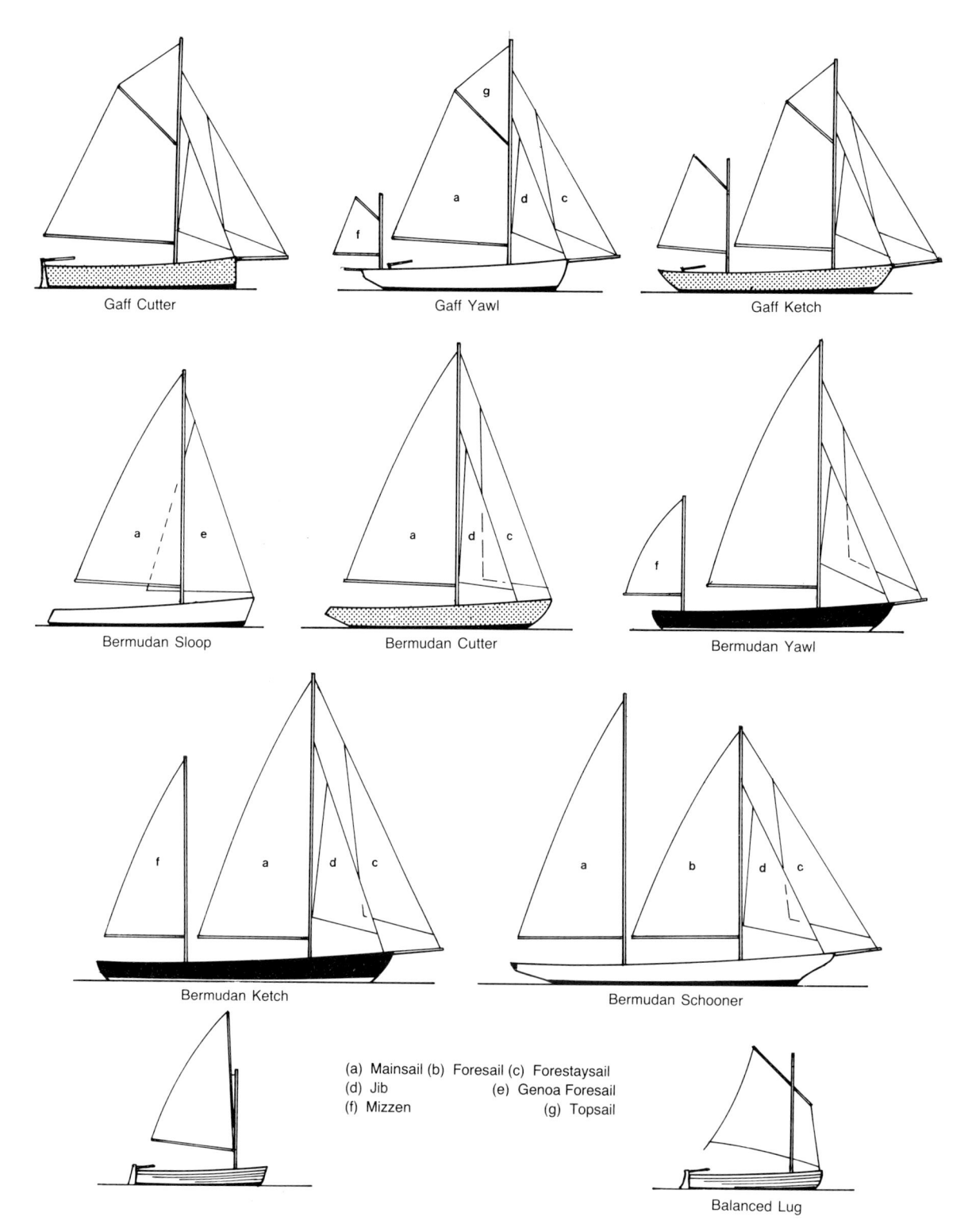

(a) Mainsail (b) Foresail (c) Forestaysail
(d) Jib (e) Genoa Foresail
(f) Mizzen (g) Topsail

Balanced Lug

Fig. 6.3 The pleasure yacht has a variety of rigs, but four distinct categories do emerge. A selection of such rigs is shown above.

and also a *mizzen* for a ketch or yawl, with a large, overlapping *genoa* for modern rigs in lighter winds. For off-the-wind there are *spinnakers* (balloon-shaped sails), *tallboys*, *big boys*, *mizzen staysails* and *spinnaker staysails*, and for heavier winds, *trysails* and *storm jibs*. The sail forward of a mast is often called a *jib*, while the forward-most jib should correctly be called the *forestaysail*, being set from the fore-stay. There is also the *lug-sail*, a quadrilateral sail lacking a boom and having the foot larger than the head, bent to a yard hanging obliquely on the mast (see Fig. 6.2).

Considering yacht rigs *per se*, the fore-and-aft rig is a comparatively recent invention and has nearly disappeared from the marine trade scene, though it is preserved in pleasure yachting. A square rig is a rare sight these days. Broadly speaking, the choice is between a Bermudan or gaff rig; the former dominates the pleasure boat scene, though the gaff rig is still loyally adhered to by traditionalists with heavy displacement boats. Incidentally, the Bermudan rig is called the Marconi rig in the USA because its complex rigging resembles a radio mast.

The pleasure yacht has a variety of rigs, many of them hybrids impossible to fit into any category. But four distinct arrangements do emerge, these being the gaff cutter and sloop, and the Bermudan cutter and sloop. The gaff-rigged sloop with single headsail is usually a smaller boat. The schooner, ketch and yawl are two-masted yachts and two examples of their rig are the gaff-rigged yawl (which may also be Bermudan-rigged) and the Bermudan-rigged ketch. There may also be modifications of the above rigs such as the single sail cat-boat of North America, the balanced lug and the wishbone ketch. A selection of such rigs and sails are shown in Fig. 6.3, together with nomenclature.

Discussing the merits of the above rigs, it must first be stated that a rig needs to suit the hull type. One should also bear in mind that one man can reasonably be expected to handle only about 500 sq.ft or about 45 m^2 of sail area. One more point: racing rules and classes may dictate the type of rig.

We now discuss the Bermudan type rigs, referring each time to Fig. 6.3. The Bermudan sloop rig, with one large mainsail and a single headsail, is the simplest arrangement and easiest to handle. It also enables the cabin to be larger because of the forward position of the mast. For production cruising keelboats of 9 metres (about 30′) and below, this type of rig predominates – although there are alternatives – and for racing keelboats it is almost universal. This type of rig is very efficient and cheap, and least expensive of all the variants is the masthead sloop. Nevertheless, for larger yachts the forward position of the mast tends to make them plunge into head-on seas and does present difficulties when heaving-to.

The Bermudan cutter rig is the most efficient, especially to windward, and is also handy and fast. But one disadvantage of this rig for large yachts is the size of the mainsail.

The Bermudan yawl rig is a compromise of the problems set up by the cutter rig's large mainsail and is popular with large cruising yachts and ocean cruisers.

The Bermudan schooner rig is extremely popular in the USA and is a good offshore sailing arrangement. Again, accommodation is improved because of the mast's forward position, but the yacht is not as fast to windward as a cutter. On the other hand, it does make up a lot off the wind on or abaft the beam.

The Bermudan ketch rig is yet a further compromise of the cutter's large mainsail, reducing it by making the mizzen sail larger, thus making the two sails nearer the same size. The rig lacks the power of the previous two types in windward conditions, but in some cases may be suitable for large cruising yachts.

Using the above as a guide to sail and rig requirements, your design should eventually be under canvas and set to go further into the creation of your own yacht.

7

SEA HORSES
Selection of an Engine

The first revolutionary step in propelling vessels through the seas was when sail took over from muscle-power and oars. Thus, the winds were harnessed. Then after thousands of years the energy within fossil fuels such as coal and oil was tapped and put under the control of steam engines. In quick succession these were followed by the reciprocating engine, turbines, and now, even water-jet propulsion for certain vessels. Who knows, the next generation of marine powerplants could very well be nuclear-based! Some nuclear-powered vessels already sail over, and under the seas. But for pleasure craft the two major types of sea horses that drive them are of the internal combustion type. They may be *Otto* cycle, burning petrol or gasoline, or *diesel* cycle, burning diesel oil fuel. Either type may have two- or four-stroke cycles.

Modern marine powerplants have developed over a comparatively recent period of time. The steam engine began to be used in ships after James Watt invented the condenser, which allowed piston operation to be by steam pressure. The next major step was when the Swede, de Laval, designed the first usable turbine and Sir Charles Parsons built the first turbine-powered steam yacht, *Turbinia*. Rudolf Diesel (1858–1913) added to the marine power story with his invention of the engine that takes his name, the diesel engine.

The diesel engine is divided into three general classes: lightweight or automotive (high-speed), medium-duty (medium-speed), and heavy-duty (low-speed). Lightweight diesels (from 100 to 3,000 bhp) have weight:power ratios from 12 to 3.5 lb/hp. In larger sizes (above 500 bhp) they are used for high-performance craft. Smaller sizes are used to propel yachts as well as other craft; they turn at about 800 to 2,000 rpm and require reverse reduction gears. Medium-duty diesels turn at 400 to 800 rpm at 35 to 10 lb/hp, with powers ranging from 800 to 4,000 bhp, while heavy-duty diesels include the large diesel propulsion units used for merchant ships, not applicable to our design.

While diesel engines have many heavy, reciprocating mechanical parts, they are very commonly used in craft under 150 ft. Petrol or gasoline engines owe much of their development to the car industry and are popular as an auxiliary power in sailing yachts. Their lighter weight per horsepower, and lower first cost, make them preferred for yacht installations. They are fitted in light pleasure craft of the runabout type and a few other larger power yachts, as well as playing an important role in high-speed craft. Nevertheless, for normal service the presence of large quantities of petrol or gasoline on board is usually unacceptable, which is one reason why diesels are more popular, despite their high cost.

The chief considerations in engine choice, once power requirements are established, are: weight and space available for installation, nature of service, number of engines to be installed, and cruising range. We also have to consider types of engine with respect to whether they are purely inboard, inboard/outboard, or purely outboard. The inboard engine takes up hull space and requires structural support and a shaft through to the stern. The inboard/outboard is a compromise between the inboard and outboard, having an inboard engine bolted to the inside of the transom, with a vertical shaft on the outside. The outboard engine is the simplest of all marine powerplants, being clamped to the outside of the transom and having a

Fig. 7.1 An inboard engine is installed in the hull and requires structural support and a shaft through to the stern. A Perkins 4.236(M) four-cylinder marine diesel engine suitable for yachts and motor vessels is shown above. (*Courtesy Perkins Engines.*)

Fig. 7.2 The inboard/outboard, as shown above, is a compromise between the purely inboard and outboard engine. (*Courtesy Perkins Engines.*)

vertical shaft to the prop. Many such units are able to swing up if they hit an underwater obstacle. Both the inboard/outboard and purely outboard require local strengthening of the hull in the form of a doubling pad or similar. The three types of engine are illustrated in Figs. 7.1, 7.2 and 7.3.

One major factor in selecting which of the above engine types is suitable to our design is available space and installation. At one time I was working on a powerboat design which was to be entered in an offshore powerboat race that has since become an annual event on the British and international scene. Its owner has also become a major competitor in this, and other events. The boat, built of GRP, was to be powered by twin marinised engines from a leading US car manufacturer and capable of making the boat fly as if it had wings. Yet on the plans their dimensions did not seem unduly large. When they were installed I went down to the boat to have a look at them. There they lay, tightly cocooned in their eggshell hull, with what seemed nary the space of a hand 'tween engine casing and shell. Yet the cleverness of design of machinery installation permitted all important parts to be reached.

The power requirements for propulsion may be covered by one, two or even more engines, as well as a combination of single engine with low power auxiliary for low speed operation. The advantages of a single engine are that, for a certain power, it is lighter than two of an equivalent type, and easier and cheaper to instal and maintain, as well as offering simpler installation,

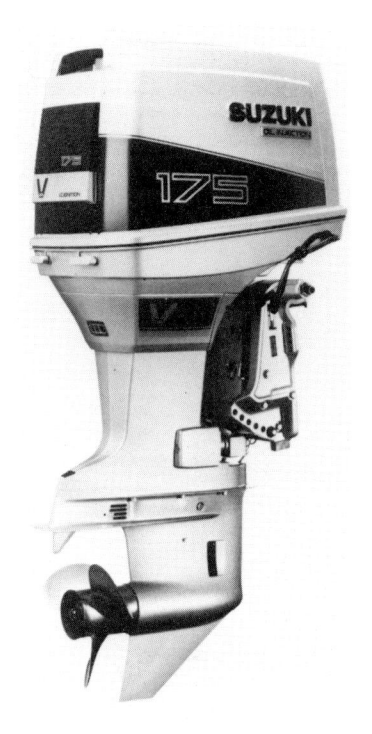

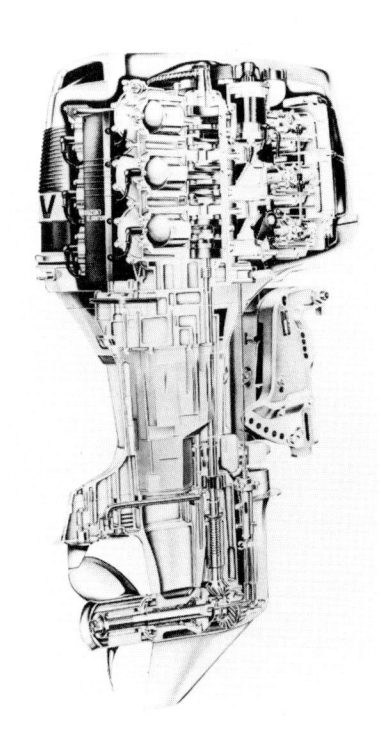

Fig. 7.3 The ouboard engine is the simplest of all powerplants for yachts and motor boats. Two Suzuki engines suitable for such craft are shown above. (*Courtesy Suzuki.*)

better protection of prop and economical use of power. Twin engines offer greater safety offshore, better manoeuvrability, and an allowance of lighter draught, while one engine only may be operated for low speeds.

Because initially one requires to find an engine whose power:weight ratio satisfies design requirements for power, weight, and space, the choice usually resolves itself between fast- or slow-turning engines. Light, high-speed craft are very sensitive to the power:weight ratio of their machinery. For heavy displacement yachts the power:weight ratio is not so critical and the choice between fast- and slow-turning may be decided by service, special needs, or merely custom and prejudice.

Speed reduction is obtained by using reduction gears, in the smaller units these being of the single-helical mating type. A very compact and lightweight unit for high powers is the *locked-train* type of double reduction gear found in modern naval vessels. Selection of gear reduction ratios is to be found in Chapter 16, and it may also be added that, for smaller engines, there are reversing gears of the disc-and-cone-type friction clutches. These gear arrangements need to be considered with the type of drive, which may be direct, vee, Z, or other. Some drive arrangements are illustrated in Fig. 7.4.

Sea Horses

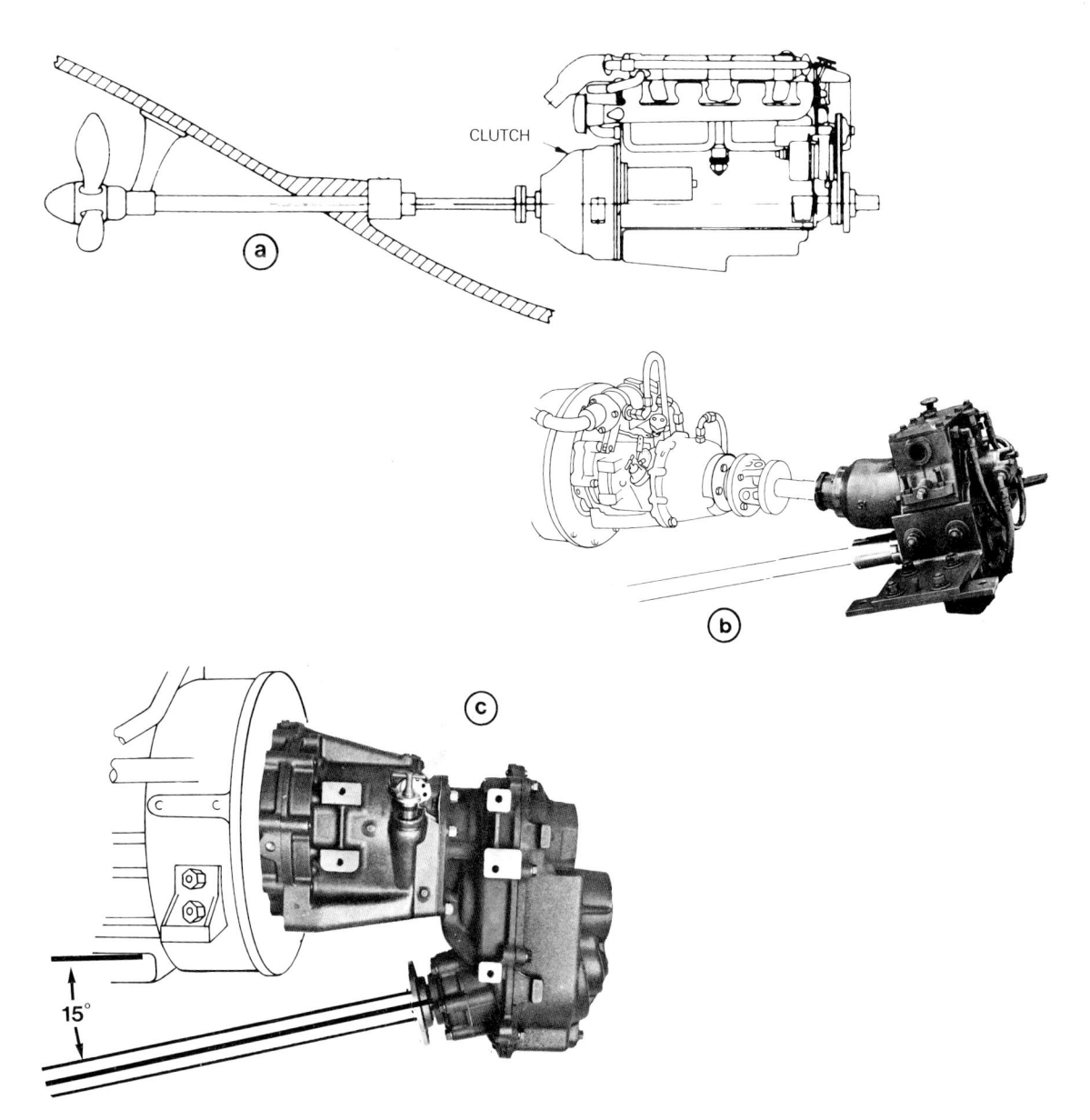

Fig. 7.4 Drive arrangements can be direct, vee, or other, depending on design considerations and suitability. Some drive arrangements are shown above. (*Courtesy Perkins Engines.*)

We now have an awareness of the practical requirements of powering. A later chapter discusses the theory and estimation of horsepower, but before this we must discuss some other elements of yacht design, such as outfit and fittings, hull and superstructure contours, approaches to, and writing the specification of design and discussing basic aids to the calculations.

8

FITTINGS AND ABLE EQUIPMENT
Discussing the Outfit and Fittings of a Yacht

If ever there was a business that touches upon every other trade, it is the marine business; if ever a designer had to know much about a little and a little about a lot, it is the marine and yacht designer. He has to be able to select, locate and install a large variety of diverse fittings and equipment and design functional systems around them. One moment he may have to design an intricate pipe system or machinery arrangement, the next he may be planning how to put a semblance of comfort and décor into cramped cabin quarters. The job of fitting out a yacht and selecting able equipment is, indeed, a varied one.

Plate 9 A most important innovation in the history of navigation and sail was the compass. The above compass, held by Elmer Sperry, is an early instrument of Sperry Marine, who now produce highly sophisticated modern counterparts. (*Courtesy Sperry Marine.*)

Fittings and Able Equipment

In the past it was just as diverse, although the equipment was more basic. There were all the appurtenances of navigation, steering and sails; they were crude and much less sophisticated than present-day counterparts, but they performed their tasks in rudimentary fashion. On the navigation side the most important innovation in the history of sail was the compass, a primitive device at the time, and of unknown origin. Other navigation aids were the Hadley quadrant, constructed by Hadley in 1731, and the sextant made by a British naval officer named Campbell in 1757.

Deck machinery such as windlasses, capstans and winches with antiquated worm gears graced the decks of those solid wooden hulls, while above there were all forms of braces, ropes, blocks and tackle. Safety-conscious yachtsmen of today would boggle at the sight of brick fireplaces – sometimes with no chimney stack – housed in the dark and smelly galleys of those past sailing ships, while communication was just a degree above Indian smoke signals, consisting of flags, whistles, the hourglass and ship's bell.

Modern marine equipment is much more advanced than when I first started my training; in fact, so much more advanced that in many cases a black box will do the job for you. It is also much more diverse and complex

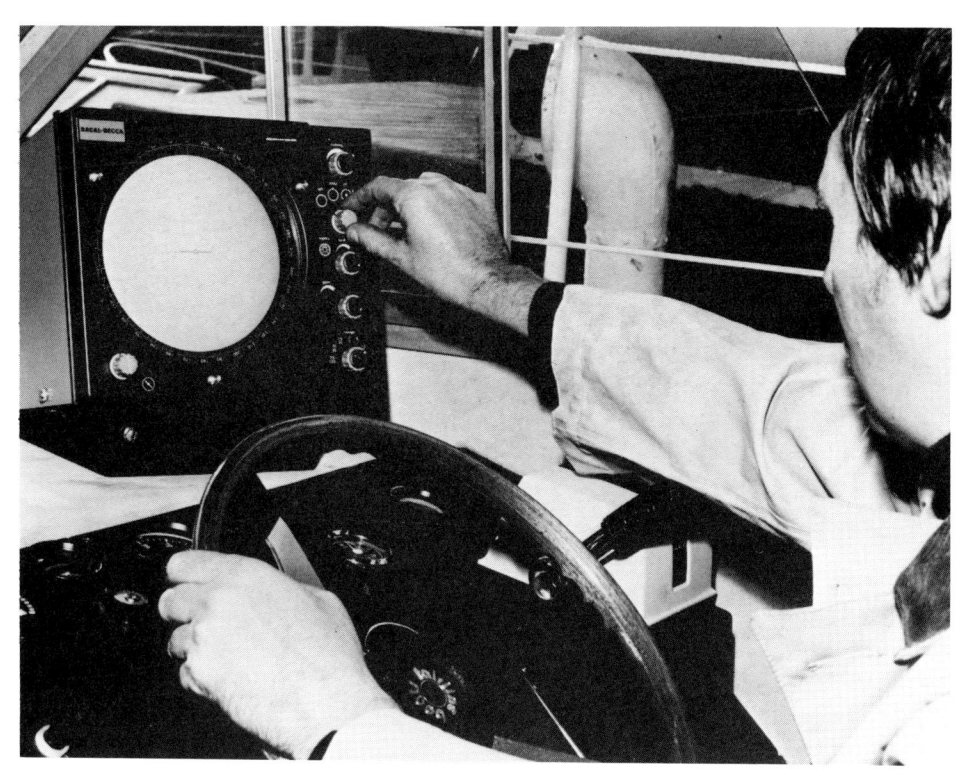

Plate 10 There are many sophisticated navigational aids for yachts and other craft on the market, including the Racal-Decca Display Control unit shown above. (*Courtesy Racal-Decca.*)

47

and this chapter can really only list the various areas of fittings and equipment in order to bring to mind the items to consider in this self-design exercise. The rest is up to you, the reader, to look up and obtain brochures and catalogues from the many manufacturers that deal in yacht and boat equipment. These will provide detailed information that will assist the design and specification.

Considering navigational aids, if you contemplate a design for offshore or ocean-going service then a radar is almost a must. A number of units are on the market (illustrations from some major manufacturers are shown in this chapter), and sophisticated versions can provide anti-collision information as well as true and relative motion of the yacht. Other navigational considerations are gyrocompasses, speed logs, depth sounders, etc., while there is a range of satellite or other navigators. Radio equipment also needs to be thought upon, as well as the basic navigational aids such as charts, sextants, etc.

Top of the list of priorities are the items connected with the sails and rigging; such items as masts, spars, fibre and wire ropes and their connections, not forgetting the actual sails themselves. We need to consider the cleats, turnbuckles and other aspects such as blocks, sheaves, etc. A selection of such fittings is shown in Fig. 8.1.

Steering gear for a yacht may be by a simple tiller or else a wheel, and a connecting system such as rod-and-link, cable, or even hydraulic or hydro-electric. Again, a number of manufacturers specialise in this area and proper selection can be made by studying their brochures. In the process, the rudder stock, pintles and bearings should also be taken into account.

Deck equipment requires thought on windlasses, winches, anchor and chain (if any), bollards, and mooring ropes. Other items for consideration are hatches, handrails, pulpits, fenders, rubbing strakes, etc. One may also consider steps and ladders and other less obvious items within the consideration of deck equipment and fittings, together with anti-skid devices or coatings for the deck and the requirements for removing any water that is shipped on board.

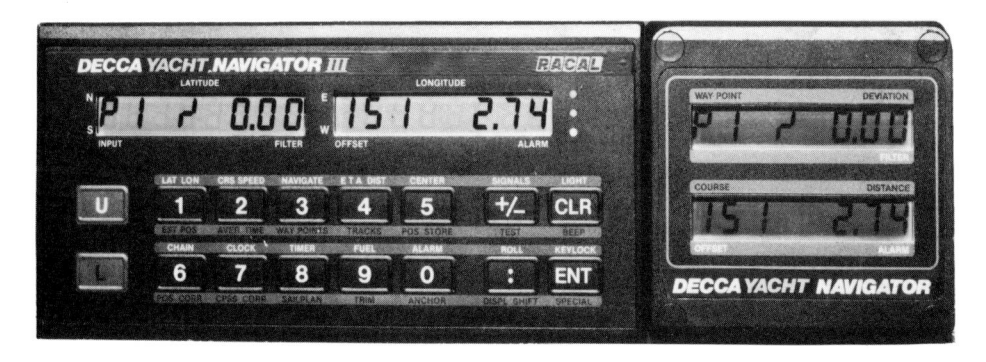

Plate 11 Another navigation aid is the radio navigation system such as the Decca Yacht Navigator III above. (*Courtesy Racal-Decca.*)

Fittings and Able Equipment

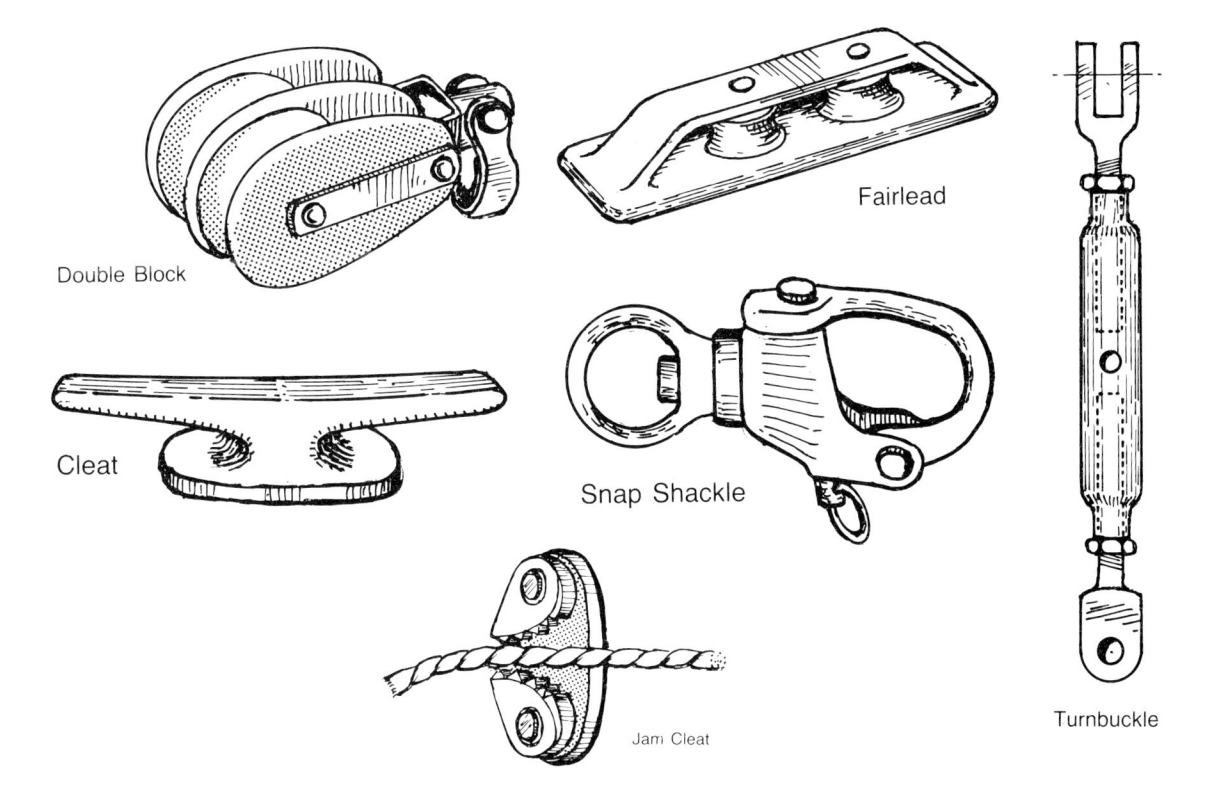

Fig. 8.1 Items to be considered in your yacht design fittings are those connected with sails and rigging, such as the selection illustrated above.

For safety and security, large yachts will carry lifeboats and davits – or else a raft or dinghy – and staple requirements will be lifebelts and lifejackets. To combat fire a system of fire extinguishers may need to be installed, while for luxury yachts a security system may be considered.

The lighting and heating requirements of a yacht can be extremely indulgent or positively spartan, depending on the economics of the design. For natural lighting the frames and material of bridge and cabin windows and portholes need to be considered, while for artificial lighting the necessary wiring, plugs, sockets, etc., are part of the equipment schedule. The power source for artificial lighting is most important, whether it be tapped from an inboard engine or from batteries, and it also affects heating requirements. With an abundant supply of energy from mechanical or electrical sources a form of central heating can be installed; otherwise, portable heaters are the only alternative.

Ventilation is yet another aspect of fittings and equipment. A forced system will require fans and a power source, but in any case the inlets and outlets of ventilation require vents, grilles and associated equipment. A selection of typical vents is presented in Fig. 8.2. One other consideration should be the ducting system to supply air to the various spaces.

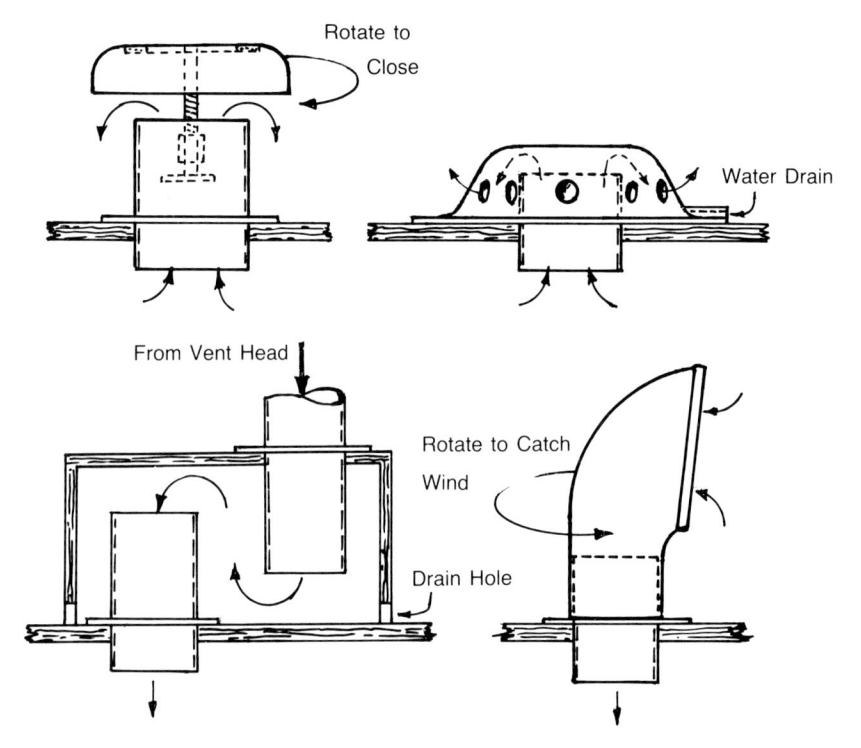

Fig. 8.2 Ventilation of a yacht requires vents, as typified by the sample shown above.

Looking at accommodation and living spaces, we need to consider cabin linings and furniture such as tables, berths, lockers, drawers and their fittings. Cooking facilities may range from a proper oven, fridge or ice-box and other modern-day household aids for luxury designs, down to a Calor gas stove or less for basic yachts. Storage space such as cupboards and drawers, should be so designed to ensure that crockery and culinary equipment are secure in rough weather. Similarly, toilet facilities can be up-market and include bath and shower, or else a very primitive form of chemical basin. A variety of such facilities is available on the market, suitable to a marine environment.

Lastly, we need to consider less tangible items of equipment such as fuel and water systems, their storage, food and stores, maintenance and repair tools, etc. Hull protection in terms of paints and coatings against corrosion and fouling are important considerations, while on a lesser level such items as clothing and utensils, lamps or candles and so on may be given thought. It seems an endless list, but it can be quite fun studying the various brochures, or when visiting a boat exhibition.

Now on to the next stretch of water in our DIY design.

9
LOG OF DESIGN
The Specification

A designer colleague of mine – who was very good at detailed design but became somewhat tongue-tied when consulting prospective owners on initial concepts – once had to draw a proposal for a motor yacht based on the *verbal* specification of his client. He meticulously executed the project and then presented it for approval.

Viewing the plan, his client – parodying that famous Nelson line '*I see no ships*' – retorted, 'But I see no chamber for nature's functions.'

Nature's functions? Suddenly light dawned.

'Oh – a *toilet*! But you didn't mention it in our talk. I thought you had some makeshift arrangement in mind.'

'Such as using the boat side,' replied the client sarcastically. 'I thought you would realise I required something more . . . *refined*!'

The problem was solved by squeezing a chamber between a shortened berth and relocated bulkhead. So the above story points out two morals of yacht design; always write a specification, incorporating as many of your ideas and intentions as possible, while at the same time appreciating that there is room for modifications and adjustments to your initial design concepts.

Large craft have extremely comprehensive specifications which run to many pages. In fact, one luxury motor yacht I once worked on listed, among other items, the owner's bedroom to be equipped with a French *Renaissance* handbasin, matching taps and *chamberpot*! We needn't be quite so detailed in our specification, but we should have a good idea of our major design requirements, as well as broad details of equipment and furnishings.

Even the wooden ships of old had a written log of requirements. For instance, a 74-gun Third-Rate ship of about 1,700 tons required 2,000 oak trees – plus other timber – to construct, and the list mentions spars, rigging and sails to be interchangeable with similar vessels. The tally for these ships would run into hundreds of blocks and sheaves, an immense yardage of sail and numerous coils of rope. There would also be guns, shot and powder, while the supply side would request salted beef, fresh vegetables, hard tack and that staple essential of a seaman's life – grog.

For our design we need to consider the following major points: its purpose, the type of hull, and how it is to be powered.

HEAVY DISPLACEMENT **LIGHT DISPLACEMENT**

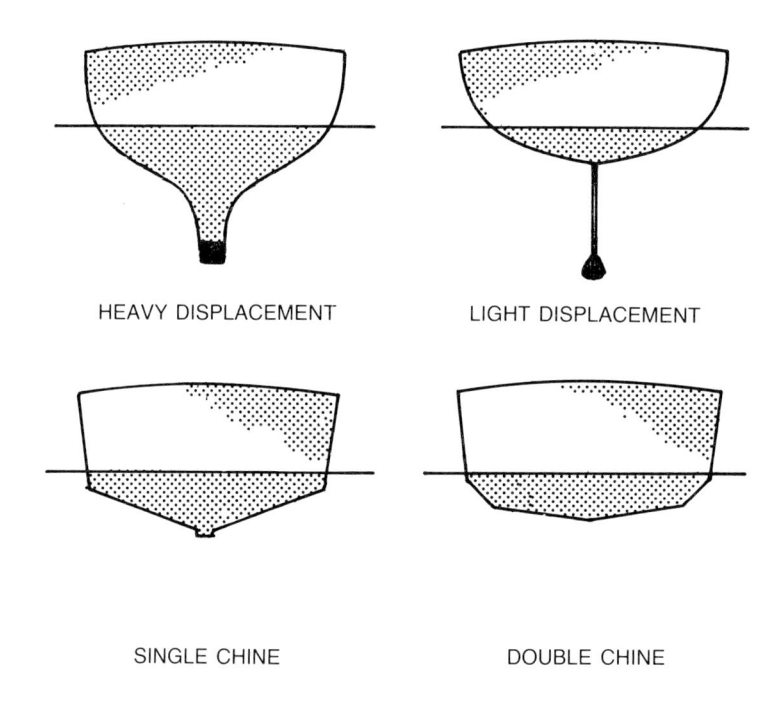

SINGLE CHINE **DOUBLE CHINE**

Fig. 9.1 A hull can be of the displacement type (light or deep), hard-chine or of planing form.

Purpose

Will the yacht be used for racing or pleasure? Is it to ply the oceans, be used for offshore sailing, or just inland waterways? Is it for day use or longer (which means berths)? Finally, will the vessel carry a large complement or merely one or two bodies?

Every one of these decisions will determine the size and type of hull, as well as its powering. And from such considerations will stream a flow of lesser ones, all leading to a complete record of your design ideas.

Hull

A hull can be of the displacement type, *light* or *deep* (this is usually of rounded bilge construction), hard-chine (single or double), or of a planing form (very flat bottomed) (see Fig. 9.1). It is obvious that the purpose or service of your craft influences the form to be selected. For example, a planing form is not suitable for a cruising yacht, and a heavy displacement vessel will hardly streak to success in competitive races. Nevertheless, there is a certain trade-off between shapes, as well as practical decisions regarding feasibility of construction, expense, etc. Obviously, the above factors will also influence the dimensions of the yacht, as well as sail area, engine power and other factors.

Powering

This is a crucial aspect of yacht design and the first decision is whether it is to be a pure sailing craft, engine-powered such as a pleasure cruiser, or be a combination of both. In a number of designs the decision is obvious, but for a yacht that might require auxiliary power a number of considerations have to be taken into account. We will discuss powering in detail in later chapters.

Plate 12 A crucial aspect of decision making in yacht design is whether it is to have engine power or be a pure sailer, such as Cougar's 65′ 0″ *Swagman*, above. (*Courtesy Cougar Marine.*)

Plate 13 Referring to manufacturers' brochures and yacht magazines will assist you in choosing and writing into the design log your deck and rigging fittings such as the sheet jammer above by Offshore Instruments. (*Courtesy Offshore Instruments.*)

Materials

The four major materials of yacht and small craft construction are mild steel, aluminium alloy, wood and glass reinforced plastic (GRP). In your choice you will have to very seriously consider the following points: ease of construction, weight and strength, maintenance and repair, and finally, costs. The materials of design have been closely examined in an earlier chapter, and in most cases there should be clear pointers to the most suitable material for your own design.

Steering

The size and type of yacht usually determines whether it is to be steered by tiller or wheel. If the steering is to be of the wheel type then one has to consider whether the connection to the rudder is to be by a link system, hydraulic or cable. The actual rudder design is analysed in Chapter 18, but as many companies now provide standard systems of steering linkage systems, it would pay to study brochures of their products in view of whether they will suit your requirements.

Equipment and Fittings

This is, naturally, a rather nebulous area at the initial stage of design. How can one decide on every small bracket or screw at such an early stage? The simple answer is to compare your own design ideas with a similar yacht already produced, and successful in service. You will also need to refer to manufacturers' brochures and relevant yacht journals.

The practice of referring to an already successful design is a standard one – as will be seen later – so while at first sight the task of writing a detailed specification appears an onerous chore, it is greatly eased by reference to existing craft. Don't worry that you are resorting to imitative practices – *most successful designs stem from previous creations.*

10

A SPIRAL ASPIRATION

Design Procedures

It's a strange chapter title – but true for yacht and boat design. Designing any vessel is an iterative process – that is, getting ever closer to the right answer – and this process is called a design spiral because development of the design can be likened to the tracing out of a spiral. Each coil of the spiral approaches ever closer to the final design configuration.

Early naval design was a somewhat random exercise, especially as all shipping was once composed of a multitude of small craft suited to individual or geographical needs. Traditionally, such craft were *built* rather than designed. In fact, the great oak forests of Britain were the seeds of the wooden ships-of-the-line, individual oaks being selected for shape and size to suit vessel construction. Then gradually theory and science entered the scene and a more orderly progression toward ship creation began to emerge.

To the layman, any engineering project seems a baffling piece of magic between a brain and the scientific knowledge it contains. When I entered the marine field I was no different. I remember as a small boy being taken by my father – a long-serving member of the Royal Navy – to Portsmouth Harbour to view the large battleships and cruisers of the Home Fleet. Not knowing even the rudiments of naval architecture at the time, I wondered how such huge lumps of metal could float on water.

'Do they have stilts underneath, Dad?' I enquired.

He smiled and shook his head.

'Not stilts. One day at school you'll learn that it's all to do with buoyancy.'

It was a new word to me, but eventually I found out its significance in yacht design – and about a man called Archimedes who discovered this principle that apparently defied common sense.

That complex floating structure stuck in my mind for many years and when I myself became involved in the design of yachts and ships my first thought was, how did one start to put all the various jig-saw pieces together? For instance, did one put a hull round an engine – or the reverse? Should the length be fixed first, followed by beam and draught, or should one start with another dimension and then design round *that*? Although there is a lot of flexibility in the design approach, I learnt that there is a certain spiral approach to one's aspirations.

While large vessel designs describe an extremely complicated spiral pathway, this can be reduced to simple essentials – which is the purpose of

A Spiral Aspiration

this book. In fact, parts of our spiral journey can be excluded without great loss; this will be indicated in appropriate chapters. The procedures that enable us to traverse this spiral whirl into yacht creation are a mix of selection, judgement, empirical and detailed calculations, plus drawing work. So let us now begin the journey.

Referring to Fig. 10.1, the first arc of the spiral starts with an outline of requirements, or the specification previously discussed. In other words, what you want in your design. Remember, this relies – as with most marine designs – on a basis vessel and, truly speaking, most creations, whether engineering or otherwise, stem from some form of prepared ground.

Length is the most important dimension in yacht and boat design (as well as ship design) so, bearing this in mind, we progress to determining the

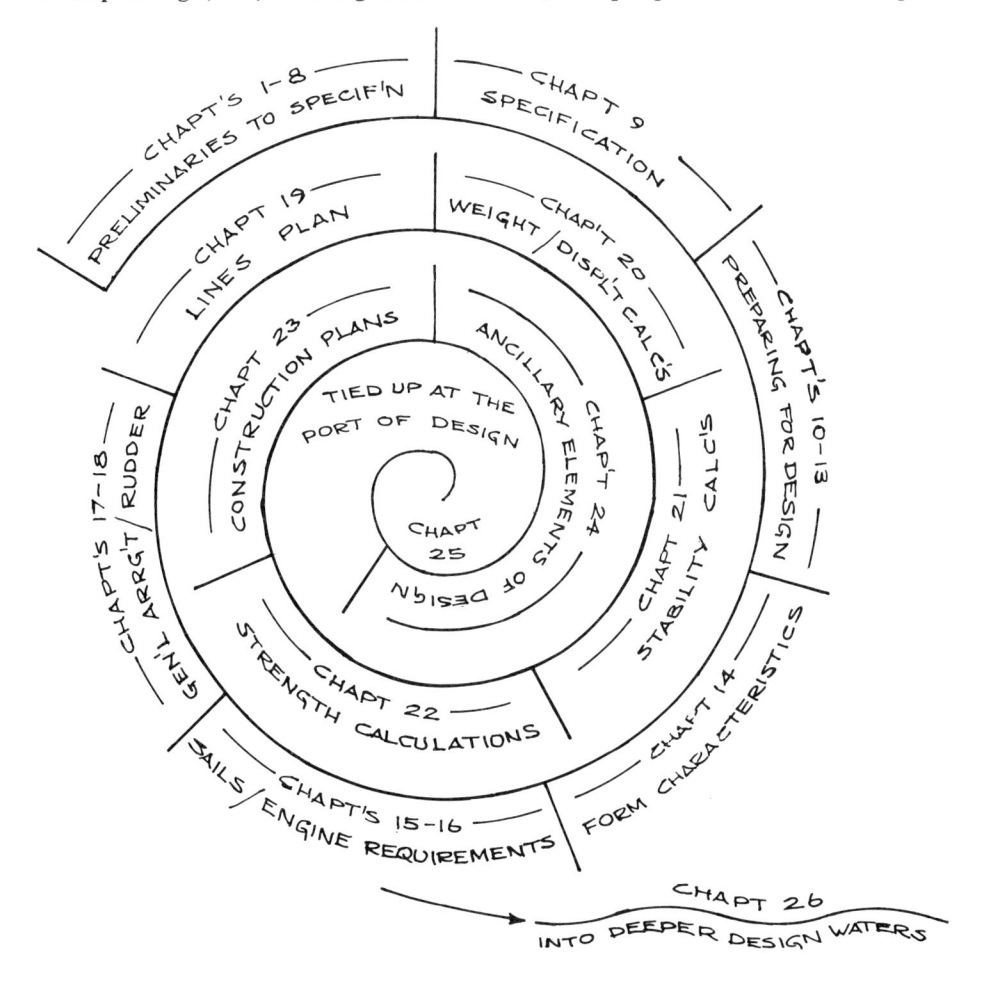

Fig. 10.1 Designing a yacht is an iterative process – i.e. getting ever closer – in spiral fashion – to our aspiration. A simple design spiral for this book is shown above.

Plate 14 Hull type and geometry, followed by sail and powering requirements, are part of the spiral journey. The Prout catamaran, above, is a sophisticated version of such considerations. (*Courtesy Prout.*)

principal dimensions and form characteristics from empirical formulae that have been arrived at by research, experience and other techniques. The formulae I have selected will be easy to calculate and from them will emerge some idea of the size, displacement, and form of yacht you desire. By the way, you should note that the dimensions obtained need to suit the waterways and harbours the intended vessel is to ply and are intrinsic to Classification Society rules, if it is to be built to such requirements.

Any form of transport requires a mode of powering. For yachts this means sails, engines (inboard or outboard), or a combination of both. We can determine the speed and horsepower at this beginning stage from various graphs and formulae – or even make an assessment from the basis design – but will later need to look more closely at this aspect of the design in terms of calculation and selection of an appropriate propeller. This can only be successfully implemented when we know the geometry of the hull.

A Spiral Aspiration

There is often debate on whether the geometry (Lines Plan) should be determined first or the layout (General Arrangement Plan) drawn. Having consulted some other naval architect colleagues (who agreed with my choice) I am going to suggest sketching a preliminary GA or layout plan first, but do remember it is a flexible decision that can be altered according to circumstances. This plan will rely heavily on the basis design and is a pure drawing exercise, refined by such various factors as judgement, experience, good practice and sheer practical limitations. There is much juggling around to be done, while a rough idea of the overall hull configuration greatly assists the task.

The Lines Plan is also a pure drawing exercise, but much more exacting than the GA, requiring far greater accuracy as well as the governing principles of judgement, experience, etc. But do not be alarmed; once into the task you will find your mind will concentrate on the essential geometric aspects and at the same time take account of the requirements demanded by your GA drawing – one good reason why the latter plan should be drawn first.

We now take our first venture into detailed design calculations – the displacement as calculated from the Lines Plan. Already some idea of the displacement and draught should have been estimated so – using the draught dimension – an accurate calculation is made, as described in Chapter 20. The calculation should hold no terrors as it requires minimal numerical skills and is presented in tabular form.

So far so good, but we now need to check whether the displacement corresponds to the actual weight of the yacht, together with machinery, equipment, stores, personnel, etc. Estimates of weights for most of the hardware on board can be obtained from brochures, catalogues or other sources, and where necessary, an intelligent guess may be made. Fuel and water in tanks can be calculated, while for more ambiguous items such as fastenings, stores, crew, etc., certain estimates and figures exist and will be given, where possible. We now come to the weight of hull and superstructure. For the preliminary stages there are certain formulae that approximate the weight of a vessel, but they are not necessarily applicable to our type of craft. At this point we are able to calculate the hull and superstructure weight, knowing the shape of design, provided that we have an idea of its structure which, for a very accurate estimate, requires the drawing of structural plans. For our purpose, at this point we can use our basis design as a guide or refer to Classification Society rules for sizes and thicknesses pertinent to our design. Any adjustments can be made later and are part of the iterative process.

The next part of our spiral journey takes us into the deeper waters of calculation. Hydrostatic data are the floating aspects of a yacht such as draught, trim and metacentre. To obtain them one has to evaluate – by calculation – the centre of gravity (cg) and centre of buoyancy (cb) of the craft. This means working out *moments* of weights, which will be explained later. The metacentre provides useful information on the stability of the vessel and the total information – at various conditions – is later presented in graph form on a drawing called the Hydrostatic Curves.

For all but very sophisticated designs in an ocean-going environment, the next step may be excluded. It investigates the stability of the yacht at various conditions, as well as sea motions and manoeuvrability. Appropriate chapters will explain the procedures and also the degree of relevance in making such calculations. You may want to pursue this line of work for the sake of knowledge and experience, even though you are quite satisfied that your vessel is very stable and seaworthy.

We now come to the strength aspects, which require strength calculations based on the weights on board. The results will determine the size of the scantlings (structure and materials), from which can then be drawn the major construction plans, followed by more detailed plans at a later stage. Some preliminary idea of strength can be derived from empirical formulae, where appropriate, and these will be presented for applicable designs. It must be said, though, that such formulae are usually more relevant to large vessels.

Having ensured the basic design is sound we enter the second coil of the spiral, where we begin to probe more deeply into each aspect of design. Large merchant vessels and highly sophisticated yachts such as those entered

Plate 15 Following the strength calculations, the construction plan and other structural drawings need to be drawn to delineate the ribs and scantlings of design, such as illustrated by the Cougar monohull above. (*Courtesy Cougar Marine.*)

for the America's Cup will usually be tested in an experimental model basin to obtain data on their hull resistance, propulsion, manouevrability, directional stability, seaworthiness, etc., and while it is highly unlikely our DIY design will go as far as model-testing (which is an expensive procedure), I have included a chapter on the subject for informative reasons. It's always nice to make knowledgeable sounds, even though you're not going to buy the instrument.

So excluding that – from our point of view – unnecessary and expensive step of tank-testing, we look back at our preliminary work on hull form and characteristics and check that machinery and outfit will fit into their allocated places, that there is enough headroom in accommodation spaces, that the relevant calculations tie in with our drawings and that we are generally satisfied all round. A number of other considerations, such as hazard conditions due to flooding, tonnage regulations and calculations and so on (which we are not too concerned with in this DIY exercise) have been included for the sake of comprehensiveness in brief descriptions at the end of the book.

Detailed structural design in accordance with Classification Society rules or calculation can now go ahead, bearing in mind that there will be a constant feedback to the initial design. This means first drawing the major construction plans, including a Midship Section, followed by outfit, machinery and propulsion (if any), electrical, steering and rudder drawings, among others, not necessarily in any set order.

The iterative process continues, referring and adjusting back and forth along the spiral, until we are happy we have a sound and seaworthy craft: a yacht conceived of our own mental labours and creation. I have made a very simple spiral (Fig. 10.1), to conform with our DIY design.

We now plunge into this creative spiral.

11

AHEAD AND ASTERN – ABOVE AND BELOW
The Contours of Design

Before entering the marine design business I had leanings toward being an artist: not the commercial type, but one who painted nubile females of Rubens sensuality, or produced Constable landscapes dripping in oils. It was a brother-in-law of mine who suggested a more feasible goal. Technical drawing and design, he suggested, demanded some artistic flair as well as scientific ability.

From such small seeds grew the roots of my marine background.

Yacht design is truly a mix of technical know-how and artistic instincts, even though the latter is greatly disciplined by the former. The marriage of the two should produce a design that is efficient, yet pleasing to the eye.

Obviously, it is the contour or profile of a yacht which has the most visual impact and this, in turn, is dictated by the lines of sheer, superstructure, bow and stern. The keel, being submerged, may not be so important, but is

Plate 16 The staid profiles of earlier yachts seem quite out of place in this streamlined age of marine, where craft like the Fairline Fifty have the appearance of a sea greyhound. (*Courtesy Fairline.*)

suitably discussed in this chapter. Remember, small yachts are quite often out of water.

The beauty of a yacht is not just in the eye of the beholder – it also depends on time and tide. Where it is to ply has a bearing, as well as the whims of fashion during the ages. The staid profiles of earlier yachts seem quite out of place in this streamlined age where economy and continuity of line predominate. (Some early yacht profiles have already been illustrated in Fig. 1.2.)

Nevertheless, the whole question of what is visually appreciable is a matter of taste, especially considering the variables in choice of sheer, superstructure, and contours, etc. A selection of shapes is given in this chapter, together with some guidelines and simple observations on their usage.

Sheer beauty (to make a pun) is not the only factor in selecting the line of deck in elevation. Initially, it is governed by the freeboard (the height of deck at amidships above the floating waterline), the headroom forward and aft, reserve of buoyancy and dryness of deck. A bold sheerline with low freeboard (see Figs. 11.1(a) and (b)) usually goes with a heavy displacement vessel and is traditionally associated with the fishing boat hull, being assumed a seaworthy feature for such craft in all weathers.

A higher freeboard is associated with light displacement vessels and provides a flatter sheerline curve (see Fig. 11.1(c)), as well as a better reserve of buoyancy. If moderate speeds are required, a yacht follows the fishing boat's example of heavy displacement and a bold sheer; otherwise, light displacement yachts and motor cruisers will be best suited to a flatter sheerline.

For motor yachts a broken sheerline (Fig. 11.1(d)) is one way of reconciling grace of appearance with space and light in cabins, while a reverse sheer (Fig. 11.1(e)) is usually associated with high-speed motor yachts. Adaptations of the sheers mentioned, such as a straight line sheer (Fig. 11.1(f)), can be implemented, where suitable, and reference should be made to your basis design.

We now consider the ends. Some bow and stern shapes are shown in Fig. 11.2 and in general, it can be said that by giving a boat overhangs fore and aft the effective waterline length is increased when sailing, which theoretically means an increase in speed. Yet again, most cruising yachts have short ends – though not too short – to make them better sea boats and lie more quietly at anchor.

Bow shape and that of the sections forward are generally considered of vital contribution to a yacht's seakeeping ability. The bold bow, like the bold sheerline, is traditionally respected for such qualities (see Fig. 11.2(a)). One popular type of sailing bow is the spoon bow (Fig. 11.2(b)). A dry and powerful bow has good breadth at the deck near the stem and the modern sloping stem, or modified clipper bow and the normal clipper bow, provide these geometric properties (illustrated in Figs. 11.2(d) and (c) respectively). Overhanging bows, illustrated in Fig. 11.2(e), serve their purpose not only

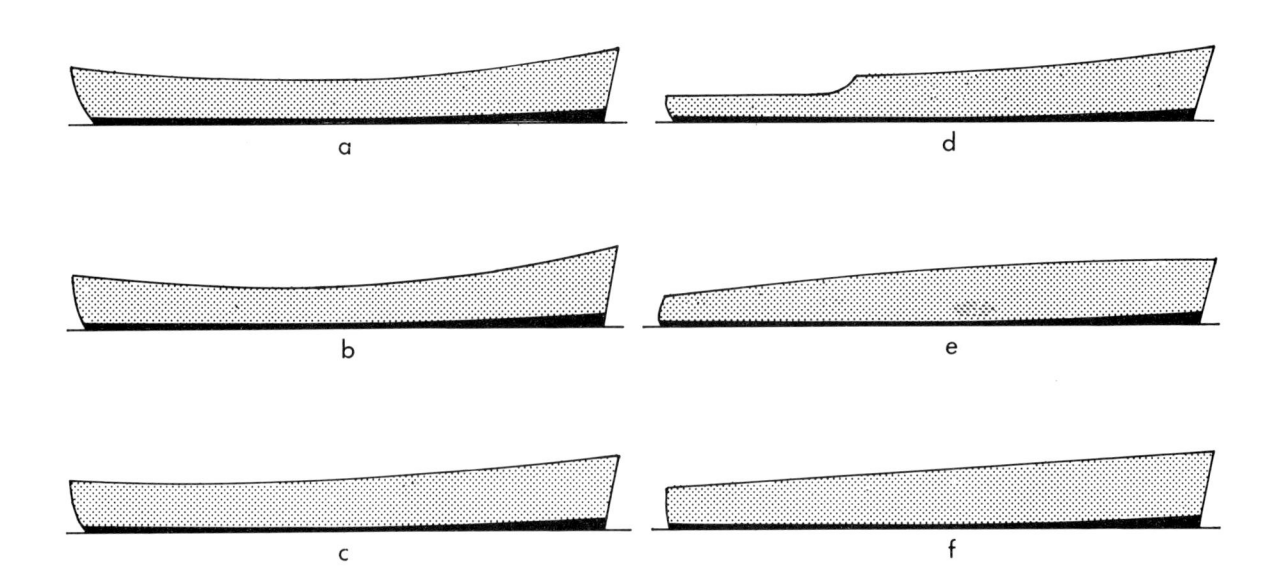

Fig. 11.1 The sheer of a yacht is dictated by a number of factors such as freeboard, headroom and dryness of deck. The sheerlines illustrated show (a) and (b), a bold sheerline with low freeboard; (c) a flatter sheerline associated with the higher freeboard of light displacement vessels; (d) a broken sheerline that reconciles grace of appearance with space and light; (e) a reverse sheer, usually for high-speed motor and racing yachts and (f), straight line sheer.

in increasing the effective waterline length (as previously mentioned), but also in enabling easing of the lines in the hull shape, enhancing the space requirements, and increasing the reserve of buoyancy. Other bow shapes have their own merits and should be considered in conjunction with the basis design selected; one type to give an old-fashioned appearance is the straight or plumb bow, Fig. 11.2(f).

The shape of stern is a somewhat theoretical choice, being initially determined by the speed:length ratio (see Chapter 12). Roughly speaking it is based as follows: Figure 11.2(g) is a cruiser stern, now not so popular, while for $V\sqrt{L}$ less than 1.34, adopt a canoe stern (see Figs. 11.2(h) and (i)) in one of its various forms. Nevertheless, other end contours also have their advantages. If other considerations enter the design – such as the need for working space at the after deck – then this may lead to the adoption of a transom stern, though an immersed type should be avoided. At higher speeds the broad, flat and immersed transom (see Figs. 11.2(j), (k) and (l)) should be adopted. Other types of stern such as counter sterns (see Figs. 11.2(m), (n) and (o)) serve their purpose according to design requirements and once again, reference is best made to the basis design.

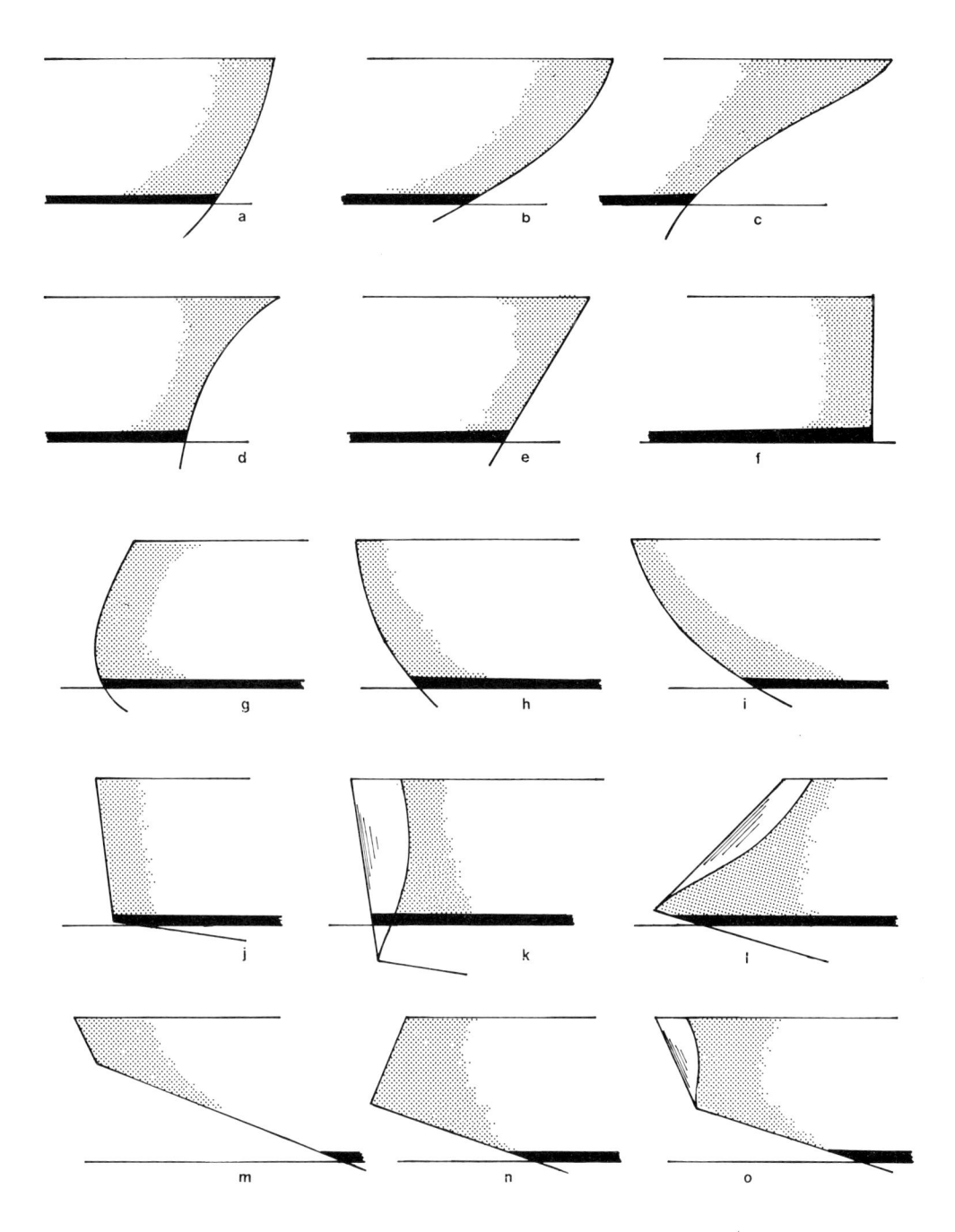

Fig. 11.2 Giving a yacht overhangs fore and aft increases the waterline length when heeled and, theoretically speaking, the speed. Other end contours also have their advantages. Figure (a) is a bold bow, (b) a spoon bow, (c) a clipper bow, (d) a modified clipper bow, (e) an overhanging bow and (f) a straight or plumb bow. Figure (g) is a cruiser stern and, (h) and (i) are canoe sterns in some of their various forms, while (j), (k) and (l) are various types of transoms for yachts and motor cruisers. Figures (m), (n) and (o) are types of counter stern.

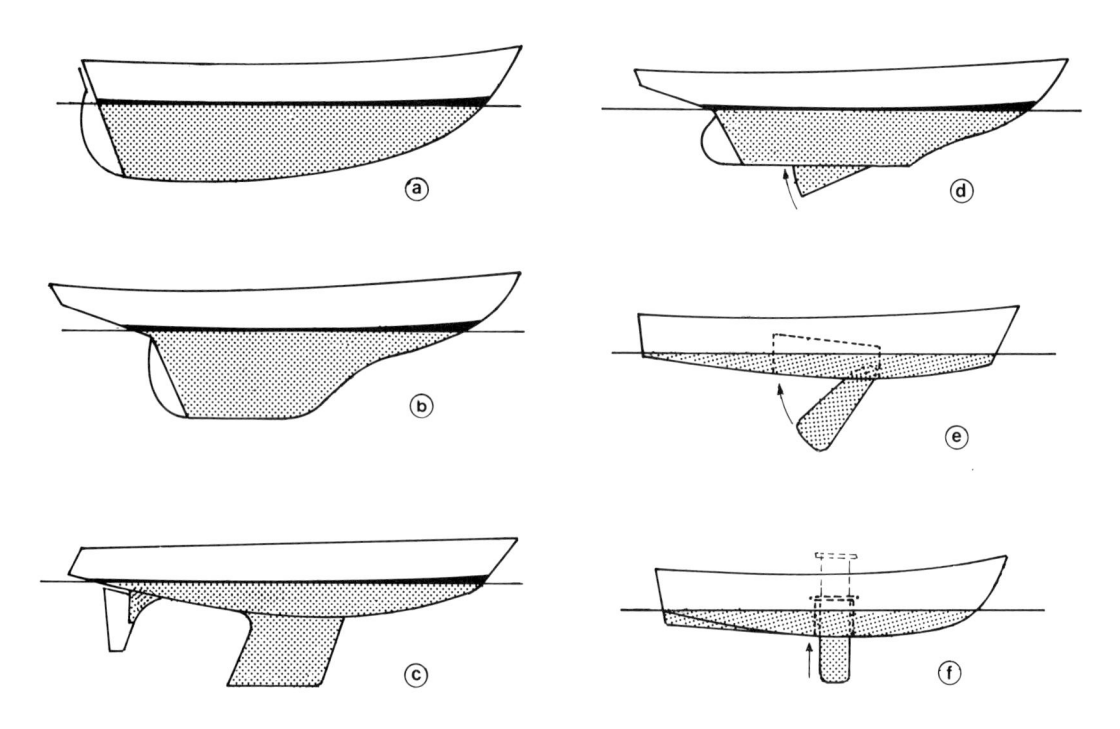

Fig. 11.3 Keels assist a yacht sailing to windward and are either ballasted or unballasted. Figure (a) is a long keel, (b) a short, deep keel, (c) a fin and skeg. Some centreboard shapes are shown in Figs (d), (e) and (f) (a daggerboard).

In the narrow sense of the word, a keelboat is a fully decked yacht, built for racing, as well as for pleasure. Obviously, all vessels have some form of keel, but for this chapter we will look mainly at keel types for sailing yachts, as well as relevant undersides of motor yachts.

Keels assist a yacht sailing to windward and are either ballasted or unballasted. The ballasted keel usually carries lead or cast iron. The main choices of hull lie between a light or heavy displacement and, for keels, between fin and skeg (Fig. 11.3c)) or a straight keel configuration. In very general terms a fin and skeg arrangement is faster. The traditional yacht of the past had a long keel (Fig. 11.3(a)) while racing boats were built with a short, deep keel (Fig. 11.3(b)). A long-keel yacht will stay on course without constant steering and is an easier boat to sail, while a short keel has less wetted surface area and thus less drag, as will be explained later. The vast majority of racing yachts – whether designed to rules or not – have a fin and skeg configuration, which greatly aids their speed and performance but does not assist directional stability (the ability to maintain a straight course). The fin keel yacht, along with other types, can also dry out easily against dock walls – a valuable asset for any yacht, though greater care should be taken than with long keels.

Ahead and Astern – Above and Below

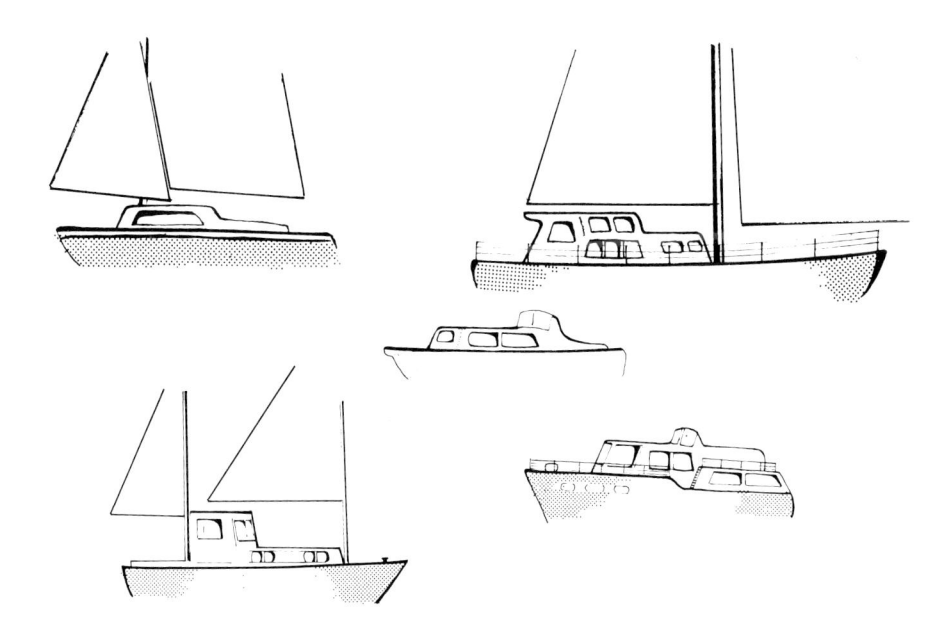

Fig. 11.4 The profile above deck lends itself most to artistic innovation in your design. There are many styles and varieties of superstructures and a selection of shapes is shown above.

Centreboard dinghies are either general day boats or racing craft. The shape of board is very important to the boat's performance and as a rule-of-thumb, try to achieve the longest and narrowest configuration possible. Some centreboard shapes are shown in Figs. 11.3(d), (e) and (f).

Conventional motor yachts and cruisers (unless they are of shallow draught and flat bottom) usually have a skeg because the hull form rises aft and also because it aids the steering and manoeuvrability of the craft. A skeg can be hollow or solid, depending on constructional aspects and materials choice. Another important advantage is that they damp out rolling.

Finally, we consider the profile above deck – the part that probably lends itself most to artistic innovation in your design. The basic technical considerations are headroom, light, and whether there is a bridge or wheelhouse. Bear also in mind that structures above the deck should offer least resistance to wind and air, which calls for some form of streamlining. One last point: a top-heavy superstructure will adversely affect stability, especially if it offers a large area to a wind blowing in the transverse direction. Because there are so many styles of superstructure and varieties of yacht, it is only possible to show a selection of shapes, which are illustrated in Fig. 11.4.

It is in the contours of design just discussed that you will most effectively express the artistic elements of your creation. The following chapters will provide the technical back-up that makes it not only a thing of beauty, but also one of efficiency.

12

NAVIGATING THE CALCULATIONS
Basic Calculations of Design

One always has a sinking feeling in the pit of one's stomach when presented with highly erudite textbook theories. I can appreciate this because part of my career has been as a teacher of scientific subjects such as mathematics and physics. I am familiar with the pained and bewildered looks of students when confronted with a mash of mathematical formulae or a stew of scientific dicta. To continue (with apologies) the culinary analogy, I also painfully remember during my post-graduate studies being given a horrendous diet of indigestible text and symbols. The area behind my own navel not only sank like a stone – it flipped over in the process.

With these torturous memories in mind – and appreciating this is your first introduction to serious calculations within the book – I have no intention of torpedoing any tummies. In fact, I hope it will prove quite palatable to digest, and not only informative, but easy and entertaining to read.

Reverting from culinary to nautical analogies, navigating a yacht or ship over a long journey requires aids such as a chart, compass, instruments, and so on. The chart of our self-design journey is the Spiral of Design (see Chapter 10), and along that journey there are some essential calculations to perform.

Briefly, our course will take us through calculations for form characteristics, displacement, centre of buoyancy (cb) and centre of gravity (cg) of a yacht, then progress to stability, strength, sail and powering, and possibly other areas such as resistance, and ancillary topics connected with outfit and equipment. To navigate through these calculations we need certain fundamental aids. They will prove extremely useful in later chapters.

Although Archimedes discovered his famous principle of buoyancy (see Chapter 20) way back, the theoretical structure of naval architecture has only recently been established, based on the work of such people as Isambard Brunel (builder of the *Great Western*, the *Great Britain* and other major engineering projects), William Froude (founder of ship model tank experimentation techniques), and Professor Osborne Reynolds (who worked on flow through pipes). A number of others established empirical formulae, while from the field of mathematics emerged a number of useful rules that could be applied to yacht and ship design. Much of this work will appear later in the book, while the basic calculation aids are presented here. The only mathematical skills you will require are an ability to perform simple arithmetic operations or use a calculator; all else will be explained.

Navigating the Calculations

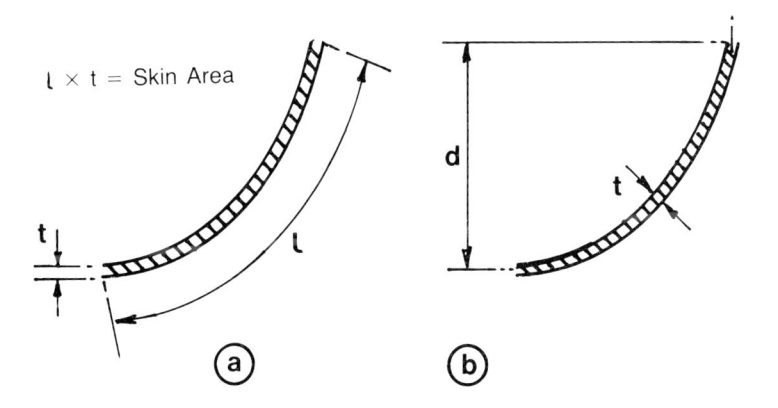

Fig. 12.1 In our yacht design we will need to find the areas (a) and other properties of curved sections of the skin of hull (b), as illustrated above. Diagram (b) refers to finding the moment of inertia of such a curve in the vertical direction, as used for the strength calculations (see Ch 22).

The main concepts we deal with at this stage are: area, volume, density, mass, weight, force pressure, stress, moments, second moments and moment of inertia. Area should present no problem to the reader, being merely the measure of the *size* of a surface or plane. In our self-design exercise we will need to find the areas of simple shapes such as rectangles, triangles and circles, but we also need to find the areas of curved shapes. A curved section of the skin of a hull (see Fig. 12.1(a)) is one example, and for this we need to determine its curved length. This can be done by running a strip of paper round the middle of its girth, then marking it off to obtain the curved length. When multiplied by the thickness (in the same units), this will give its area. Very important areas in yacht design are those of waterlines and sections. One approximate rule is illustrated in Fig. 12.2(c), together with the calculation in Table 12.1. Equidistant ordinates or stations need to be drawn along the base of curve and the assumption is that the divisions can be approximated to a rectangle by averaging the consequent ordinates.

Obviously, the above method is not very accurate – especially for sharp curves – and a more exact method is to use *Simpson's First Rule* which, because of its universal application and importance in naval architecture calculations, will be discussed in detail. A number of other rules exist to calculate the areas under curves, including some by Simpson, but it is his First rule that is most readily applicable to marine design. It is based on an *odd* number of ordinates such as the stations of a yacht or ship. All the rules provide what are called multipliers and are derived from a branch of mathematics known as *calculus*. We will begin with the basic form of Simpson's First Rule.

The basic multipliers (SMs) for Simpson's First Rule, using three ordinates (see Fig. 12.2(a)), are 1, 4, 1. This means the first ordinate is multiplied by one, the second by four and the third by one again, as shown in Table 12.2. Extending the odd number of ordinates to, say, five, means

Table 12.1

ORD No.	ORD LENGTH	AVERAGE
1	3.0	2.85
2	2.7	
3	2.0	2.35

Ordinate spacing or common interval (CI) = 3.0m

Area = 3.0 (2.85 + 2.35) = 15.6m^2

Note. If ordinate spacing were 2m we would have
2.0 (2.85 + 2.35) = 10.4m^2

Table 12.2

ORD No.	ORD LENGTH	S.M.	AREA FUNCTION
1	3.0	1	3.0
2	2.7	4	10.8
3	2.0	1	2.0
		Total =	15.8

Area = $\dfrac{15.8 \times 3.0}{3}$ = 15.8m^2

Note. If ordinate spacing were 2m we would have $\dfrac{15.8 \times 2.0}{3}$ = 10.53m^2

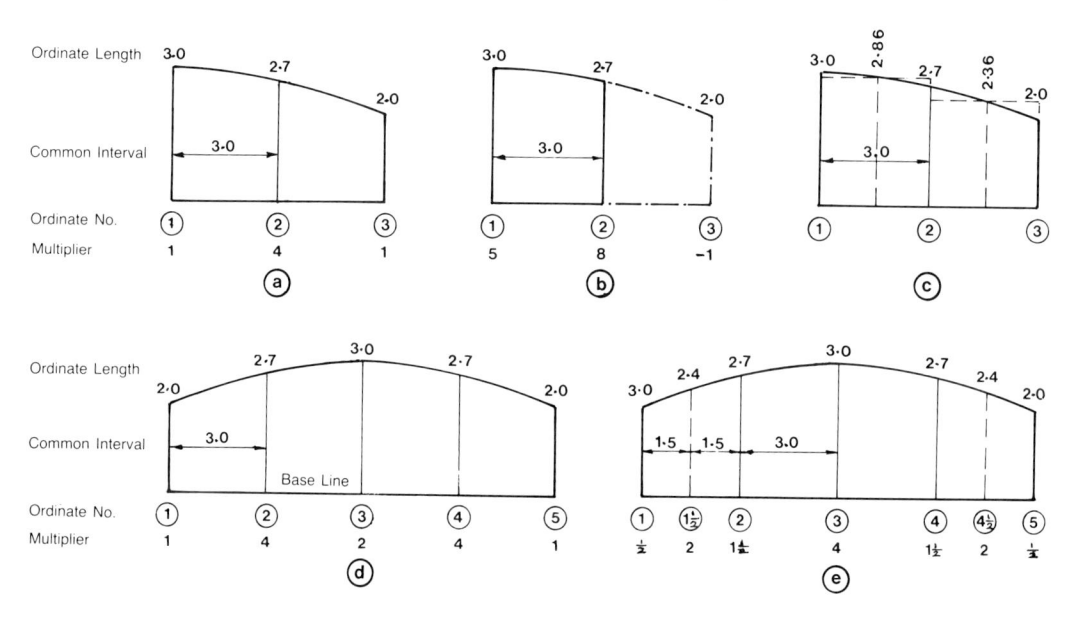

Fig. 12.2 Simpson's Rule is commonly used to find the areas and other properties of yacht curves. The above figures show how the rule may be applied – using the tables in the chapter – as well as that for a more simple approximate rule.

Table 12.3

Col 1	2	3	4	5	6	7	8	9	10	11	12
1	2.0	1	2.0	4.0	4.0	0	0	0	0	8.0	8.0
2	2.7	4	10.8	7.3	29.2	1	10.8	1	10.8	19.7	78.8
3	3.0	2	6.0	9.0	18.0	2	12.0	4	24.0	27.0	54.0
4	2.7	4	10.8	7.3	29.2	3	32.4	9	97.2	19.7	78.8
5	2.0	1	2.0	4.0	4.0	4	8.0	16	32.0	8.0	8.0
Totals			T4 = 31.6		T6 = 84.4		T8 = 63.2		T10 = 164.0		T12 = 227.6

Col 1 is Ordinate No.
Col 2 is Ordinate Length
Col 3 is S.M.
Col 4 is Area Function = Col 2 × Col 3 (Total = T4)
Col 5 is Ordinate Length^2
Col 6 is Moment Function = Col 3 × Col 5 (Total = T6)
Col 7 is Lever = No of CI's from Ord. 1
Col 8 is Moment Function = Col 4 × Col 7 (Total = T8)
Col 9 is Lever^2 = Col 7 × Col 7
Col 10 is Inertia Function = Col 4 × Col 9 (Total = T10)
Col 11 is Ordinate Length^3
Col 12 is Inertia Function = Col. 3 × Col 11 (Total = T12)

$$\text{Area} = \frac{CI \times T4}{3} = \frac{3.0 \times 31.6}{3} = 31.6 \, m^2$$

$$\text{Centre of Area (from Ord 1)} = \frac{CI \times T8}{T4} = \frac{3.0 \times 63.2}{31.6} = 6.0 \, m$$

$$\text{Centre of Area (above base)} = \frac{T6}{2 \times T4} = \frac{84.4}{2 \times 31.6} = 1.34 \, m$$

$$\text{Moment of Inertia (about Ord. 1)} = \frac{CI^3 \times T10}{3} = \frac{27 \times 164.0}{3} = 1476 \, m^4$$

$$\text{Moment of Inertia (about base)} = \frac{CI \times T12}{3 \times 3} = \frac{3.0 \times 227.6}{9} = 75.9 \, m^4$$

Table 12.4

ORD No.	ORD LENGTH	S.M.	AREA FUNCTION
1	2.0	½	1.00
1½	2.4	2	4.80
2	2.7	1½	4.05
3	3.0	4	12.00
4	2.7	1½	4.05
4½	2.4	2	4.80
5	2.0	½	1.00
		Total =	31.70

Table 12.5

ORD No.	ORD LENGTH	MULT^R	AREA FUNCTION
1	3.0	5	15.0
2	2.7	8	21.6
3	2.0	−1	− 2.0
			34.6

$$\text{Area} = \frac{34.6 \times 3}{12} = 8.65 \text{m}^2$$

Ordinate spacing or common interval (CI) = 3.0m

$$\text{Area} = \frac{31.70 \times 3.0}{3} = 31.7 \text{m}^2$$

the multipliers change slightly to give 1, 4, 2, 4, 1, as shown in Fig. 12.2(d). The resultant calculation is given in Table 12.3. The same pattern is continued for more ordinates so that for 11 (the number used in yachts) we get 1, 4, 2, 4, 2, 4, 2, 4, 2, 4, 1. Notice that in the table calculation the ordinate spacing, or *common interval* (CI) as it is normally called in marine design, has to be divided by three – in other words, one-third the common interval. Summing the length × SM values in the column (called *area function*), then multiplying by one-third this common interval, gives the area.

For greater accuracy near the ends of curves, especially the sharp ones that occur at the ends of a hull, we can modify the rule to suit. The last division at each end is sub-divided and Simpson's three ordinate rule, with the multipliers *halved*, is applied. The remaining portion in the middle uses the ordinary SMs and the result for five ordinates is shown in Fig. 12.2(e), and the calculation in Table 12.4. One last rule that is often useful is the 5,8,−1 Rule. This is illustrated in Fig. 12.2(b), and the calculation in Table 12.5.

We can also use Simpson's Rule to find the volume of curved shapes such as that of a yacht hull. In this case the ordinates are the cross-sectional areas of the shape, obtained either by the rule or other methods. In this fashion we can continue the calculation (as will be seen in later chapters) to determine centre of area, centre of buoyancy and other properties of the curved shape or volume.

We now look at some fundamentals of physics and mechanics, beginning with density, mass, weight and force. The *density* of a material is the amount of *mass* contained in a unit cube of the material. If we know the total mass of an object then we can find its density by *dividing* by its volume, i.e.

$$\text{Density} = \frac{\text{mass}}{\text{volume}}$$

On the other hand, if we know the density and volume, then the total mass is given by *multiplying* the volume by the density, i.e.

$$\text{Mass} = \text{volume} \times \text{density}$$

Similar to density, *mass* is the composition of matter within a body – all its molecules and the compactness with which they are packed within its volume. Mass is intimately connected with force and weight, as Isaac Newton discovered in his Laws of Motion. The *force* on a body, he found out, is proportional to its mass and the acceleration acting on it. This is given by the simple formula:

$$\text{Force} = \text{mass} \times \text{acceleration}$$

Weight is a funny concept – not in the humorous sense, but in the fact that it depends on where you live. Many laymen – and even engineers – confuse mass with weight. Using Newton's Law as stated before, the acceleration in this case is that due to gravity. Now the gravitational attraction of a body depends on its size and mass. A small star with a packed mass will have a much higher gravitational pull than a large, gaseous one. Our earth has six times more gravitational pull than the moon – the prime reason why astronauts can bounce on the moon's surface like ping-pong balls. The acceleration on this planet is about 10 metres per second, per second (10 m/s^2), equivalent to around 32 ft/s^2. Applying it to the formula, a body of mass 500 kg will weigh approximately $500 \times 10 = 5{,}000$ *Newtons*, the unit of weight and force in the International System of units (SI) being Newtons, or 5 tonnes force (5 tonnef) in metric units (1 tonne = 1,000kg).

Stress and *pressure* are somewhat similar, pressure being defined as the force per unit area acting normally or at right angles to a surface, and stress – also acting normally to the surface – as the force per unit area transmitted through the material of a body at rest. We will deal in more detail with pressure and stress in the buoyancy and strength chapters.

I forget who made this rather ironic claim, but one ancient scientist (possibly Galileo) said that given a long enough lever, he could shift the world. This is the essential principle of *moments*, which will frequently occur in the later calculations. A small force applied at a distance from a body, via a lever (see Fig. 12.3), will have a magnified effect equivalent to another, greater force of lesser distance. From the figure we can see that 100 kg force applied at 10 m gives the same moment as 200 kg force at 5 m, both balancing the moment of the body, which is 400 kg force at 2.5 m.

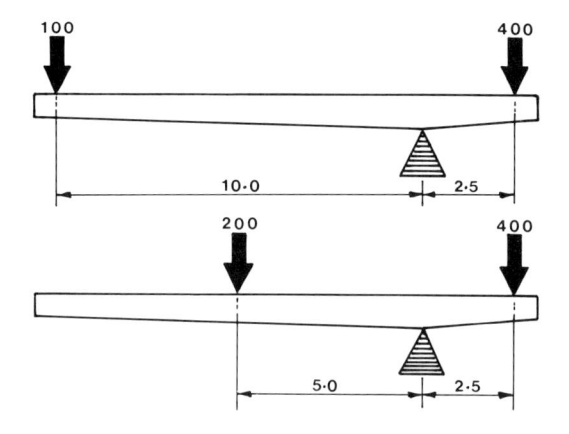

Fig. 12.3 A small force applied at a large distance is equivalent to a large force at a lesser distance, as illustrated above. This is the essential principle of moments.

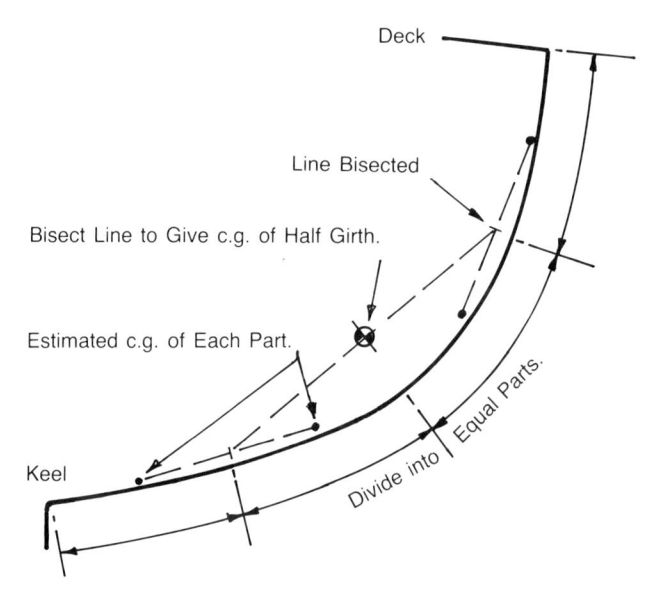

Fig. 12.4 The above illustration shows how to graphically find the centre of gravity (c.g.) of curves such as the cross-section of part of the skin of a yacht hull.

Moments are the tools used to find centres of gravity (cgs) of bodies and centres of areas and volumes. The centre of volume of the underwater part of a hull is the centre of buoyancy (cb). The cgs of simple, regular-shaped bodies such as rectangular items are in the middle, assuming a constant density all over. The cg of the curved section of, say, the skin of a hull, as illustrated in Fig. 12.4, is approximated as shown. When there is a more

Navigating the Calculations

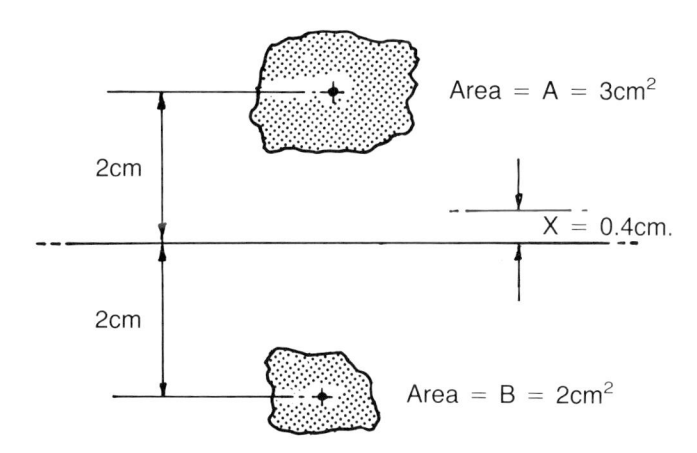

Fig. 12.5 The figure depicts how to find –, using moments – the combined centres of area or centres of gravities of bodies about an axis. This approach is used in the strength calculations of a yacht hull.

complicated structure, such as the midship section of a hull, the calculation is lengthy, but based on the simple example presented.

Suppose there are two areas A and B equidistant above and below a horizontal line, as depicted in fig. 12.5. The *true* centre of both figures is a parallel line of unknown distance X above this line. This shift X, multiplied by the combined areas A and B, must give a moment equal to the difference between the separate moments of areas. Using the numbers in the figure, we solve for the value of X in the following simple equation:

$$X \times \text{added areas of A and B} = \text{Moment of A} - \text{Moment of B}$$
$$X \times (3 + 2) = (3 \times 2) - (2 \times 2)$$
$$X \times 5 = 6 - 4$$

This works out as:

$$5X = 2$$
or
$$X = 2/5 \ (0.4 \text{ cm}) \text{ movement upwards}$$

It should be noticed that the resultant moment was divided by the total area to obtain the shift X and if both areas were above the line then we would *add* the moments to obtain the shift. From this we can state a simple formula for obtaining cg or centre of area:

$$\text{cg (or centre of area)} = \frac{\text{resultant moment}}{\text{total area}}$$

For curved shapes such as Fig. 12.2(d), using Simpson's Rule, we only

75

Design Your Own Yacht

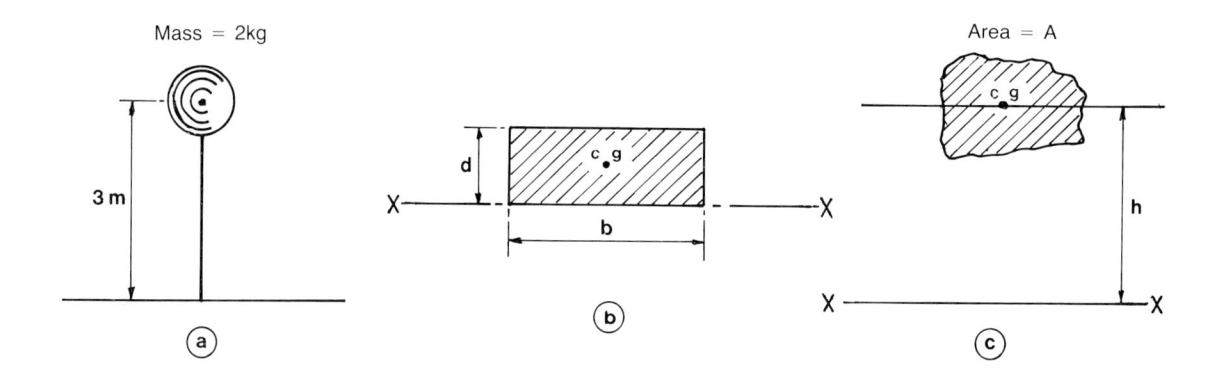

Fig. 12.6 Moments of inertia (*I*) and Radius of Gyration (*k*) are concepts that connect the resistance to motion a body offers when at rest. The Moment of Inertia of yacht waterplanes and hull cross-section are intrinsic to the stability and strength calculations.

need to use levers within Table 12.3 (Col. 7), then multiply by the common interval at the end of the calculation. It is the same as if we multiplied by the separate distances.

The calculation in the table is simply the common interval 3 multiplied by the total of column 8, then divided by the total of column 4, and does not require one-third the interval because this will cancel out for both. The result is the *horizontal* position of the centre of area from the left-hand end. But we also need to know the *vertical* position to locate it exactly.

The useful calculus shows that we have to square the ordinates (Col. 5), then put them through Simpson's multipliers to obtain column 6. The totalled result is then simply divided by the area function (total of Col. 4), and divided by two (or multiplied by a half). The answer gives the distance of centre of area above the base.

Beyond moments and centres there is then a concept in physics called *Second Moment of Area* or *Moment of Inertia* (*I*). *Inertia* is the resistance of a body to motion when at rest and *I*, in terms of mathematical explanation, is rather involved. One tangible way for you, the reader, to comprehend the meaning of *I* is to appreciate how much harder it is to set a heavy wheel turning (or stop it) than a light one. Moment of inertia *I* is a measure of such difficulties in changing the speed of turning and is bound up with what is called the *Radius of Gyration* (*k*). This radius of gyration (the distance from a certain axis to the cg of a body), when squared and then multiplied by the mass of the body, gives the moment of inertia of the body from that axis. As a simple formula:

$$I = Mk^2$$

where M = mass.

As an example, Fig. 12.6(a) shows a mass of 2 kg with a distance from an axis of 3 m to its centre. In this case I would be:

$$I = 2 \times 3^2 = 2 \times 9 = 18$$

Formulae for the I of rectangular shapes will be very useful later in the book. One is given for I through the centre of a rectangle and the other, about one side. Referring to Fig. 12.6(b):

$$I \text{ (through centre)} = \frac{\text{Area of rectangle} \times d^2}{12} = \frac{bd^3}{12}$$

The d^3 part of the formula comes from the fact that the rectangle area = $b \times d$ and when multiplied by d^2, will give bd^3.

$$I \text{ (about one side } XX) = \frac{\text{Area of rectangle} \times d^2}{3} = \frac{bd^3}{3}$$

In yacht design our axis is usually away from the rectangular or other shapes, and needs to be adjusted to suit. For this we use the following formula (see Fig. 12.6(c)):

I_{XX} = Moment of Inertia about cg + Area × square of distance from axis

$$I_{XX} = I_{cg} + Ah^2$$

The I for curved shapes such as at the bilge (see Fig. 12.1(b)) uses the virtual length approximation referred to before, then employs the above formulae.

The I for a curved area such as in Fig. 12.2(d) is next considered and without entering into detailed explanation (which is mathematically too difficult for this book) the ubiquitous calculus once again provides the tool for this. The resultant formula for I from one end shows that (referring to Table 12.3) we must square the levers (Col. 7) to obtain column 9, multiply by column 4 to obtain column 10, then halve the answers (the halving can be done at the end of calculation). Obviously, this means that the common interval must also be squared which, together with the normal common interval divided by three, means the end calculation includes the cube of the common interval.

For I about the base, another derived formula requires that we cube the ordinates (not functions), multiply them by Simpson's multipliers, as shown in columns 11 and 12 of Table 12.3, with the end calculation using one-ninth the common interval, because in the derivation from the calculus we obtain a third of the cube of ordinates. When combined with the common interval this gives one-ninth (CI/3 × ⅓ = CI/9).

For parts of yacht design calculation which relate to stability and some other aspects, a branch of mathematics known as *trigonometry* enters into the formulae and computation. Basically, there are three ratios involved, connected with a 90° or right-angled triangle. Referring to Fig. 12.7, the three ratios are given by:

$$\text{Sine (Sin) A} = \frac{a}{c}$$
$$\text{Cosine (Cos) A} = \frac{b}{c}$$
$$\text{Tangent (Tan) A} = \frac{a}{b}$$

Tables for these ratios are found in all standard mathematics and engineering texts and handbooks.

One final calculation aid of great importance in yacht design is known as the *speed:length ratio*. This ratio, already mentioned in Chapter 2, is given by:

$$\text{Speed:length ratio} = \frac{V}{\sqrt{L}}$$

where V = speed of yacht or boat in knots and L is length between perpendiculars or along the floating waterline in feet (the $\sqrt{}$ symbol means square root). As an example, suppose our design was estimated to travel at 6 knots and had a waterline length of 25 ft, then:

$$\text{Speed:length ratio} = \frac{6}{\sqrt{25}} = \frac{6}{5} = 1.20$$

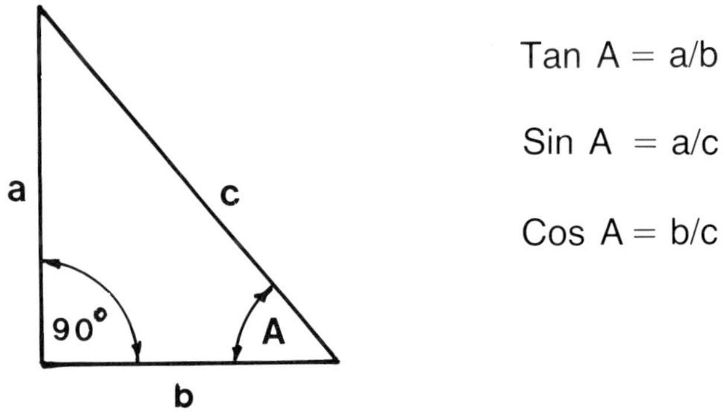

Tan A = a/b

Sin A = a/c

Cos A = b/c

Fig. 12.7 Trigonometric ratios enter into many yacht design calculations, especially stability aspects, sail and rudder forces. The above figure illustrates such ratios.

Similar to trigonometric ratios, tables of powers and roots are in all standard mathematics and engineering textbooks.

Finally we come to units. For many years in the western world we have lived with two sets of units, the British Imperial system and the metric system. When the European Common Market was formed, the metric system began to displace the British system of feet and inches, pounds and ounces used on that island. In 1960, the International Bureau of Weights and Measures (BIPM) adopted the name *International System of Units*, abbreviated to SI units, in the establishment of a comprehensive specification of units of measurement covering seven quantities, including length, mass and time. The SI units are metric in such quantities as length and mass, but nevertheless both systems still prevail in the engineering and scientific fields, especially in the USA. For marine design purposes the British units of feet, inches and tons force (tonf) and metric units of centimetres, metres, and tonnes force (tonnef) are still in common use and this book uses both units – with approximated conversions – where applicable.

In many cases it is only possible to express some empirical and other formulae in one set of units because of the way they have been derived. In that case you should work in those units then convert to your own units after finding a solution. Conversions are given at the back of book.

So, with these aids at your command, you should now navigate the calculations to a successful journey of design. *Bon voyage*!

13

AS THE WATERS SLIP BY

A Brief Explanation of Flow and Resistance Round the Hull Design

Before considering powering and sail estimates for our design, a knowledge of the flow and resistance round yacht hulls will prove very useful. It also adds to our total awareness of yacht design and yachting.

Liquids are in that nebulous state between solids and gases. For instance, apply enough heat to water and it will evaporate into steam; freeze it and the result is ice.

Most of the surface water on our planet is contained in the seas and oceans. In fact, about four-fifths of the earth's surface is covered with water. At the same time, this water surface is interfaced by a gas – air. When the sea surface is ruffled by wind or storm then waves or swell will form, depending on the strength and duration of wind plus the fetch of water over which it blows.

In our self-design we are not only concerned with the interaction between waves and hull, but also the yacht's passage through relatively calm water. The knowledge is pertinent to our form characteristics (see Chapter 14), sail forces and powering (Chapters 15 and 16), and the Lines Plan (Chapter 19). Such knowledge is not only relevant to design, but plays a part in seamanship qualities and maintenance of the hull.

During my student days the bearded lecturer who taught us marine and allied subjects took us one day to the college fluids laboratory and indicated a long, glass tube resting on supports. He went over to it, switched on a pump, and water began bubbling through at a chaotic rate.

'Take note of what happens,' he said, using a hypodermic syringe to insert some coloured fluid into the middle of flow, through an inlet at the top of the tube. The body of water immediately became dyed and streamed away.

'Take note of what?' I thought, sceptically. The whole thing seemed obvious.

The lecturer lowered the rate of flow, and then inserted the syringe again.

'Now watch this,' he said.

The students, including me, gaped in fascination as a long thread of coloured water – just like a strand of wool – progressed down the tube. A few more injections at different points in the stream and the tube began to look like a stick of rock candy.

'*That*' continued the lecturer, 'is the phenomenon known as *streamline* or *laminar* flow. It is as if the water behaves as separate laminates or pipes. Now, if I increase the flow . . .'

As the Waters Slip By

The coloured threads began to waver as he turned up the control knob, then broke down into turbulence when the water reached a certain speed.

It was a man called Osborne Reynolds who first made analytical studies of flow through tubes and pipes. He established that at a certain *critical speed* the flow changed from laminar to *transitional* (wavering), then, with increased speed, *turbulent*, flow. It is these types of water flow round a yacht hull, as illustrated in Fig. 13.1, that we are interested in. In fact, we want to know the water's effect as it slips by our hull.

In a yacht – or indeed, any other vessel – it is never possible to have purely laminar flow along the whole length of the hull, though a smooth, newly painted one achieves a close approximation to this state for quite a distance back. The aims of design are to achieve laminar flow because the turbulent kind exacts a high penalty in terms of resistance and drag.

Looking more closely at the fluid mechanics involved, believe it or not, a yacht moving through water carries with it a thin film of its watery surroundings at the waterline and below. This laminar film (shown in Fig. 13.1, but highly exaggerated) is significant with regard to the *smoothness* of hull because most of the laminar resistance is concentrated within the film. If it is disturbed or pierced by hull roughness or undue projections then turbulent flow will occur – which is to be avoided at all costs, where possible. The region of totally disturbed flow round a hull – laminar, transitional and turbulent – is known as the *boundary layer* and represents a loss of energy by the yacht (resistance or drag) as it drives through the water.

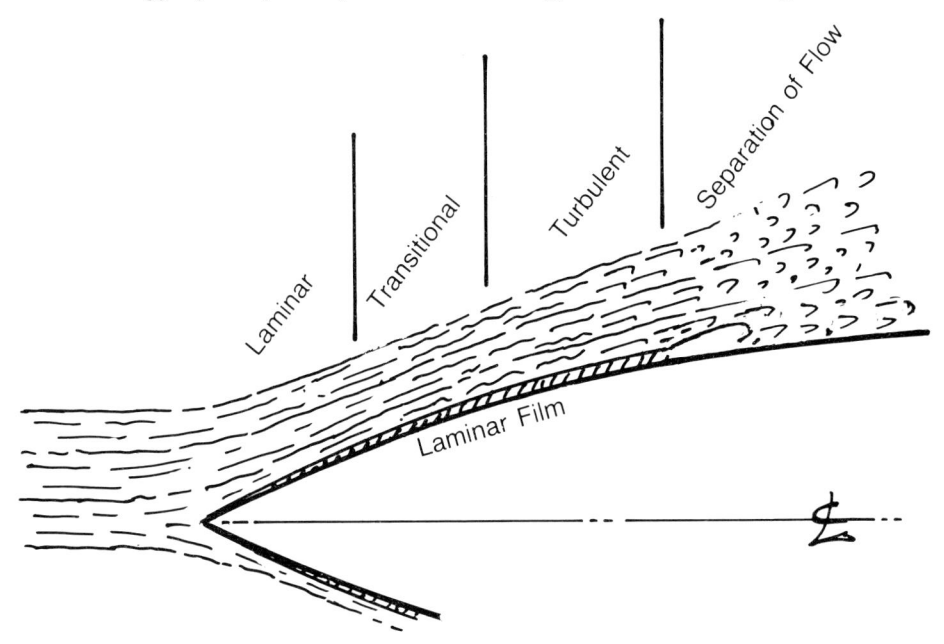

Fig. 13.1 As the waters slip by our DIY design of hull so a pattern of laminar, transitional and turbulent flow is set up, as shown above.

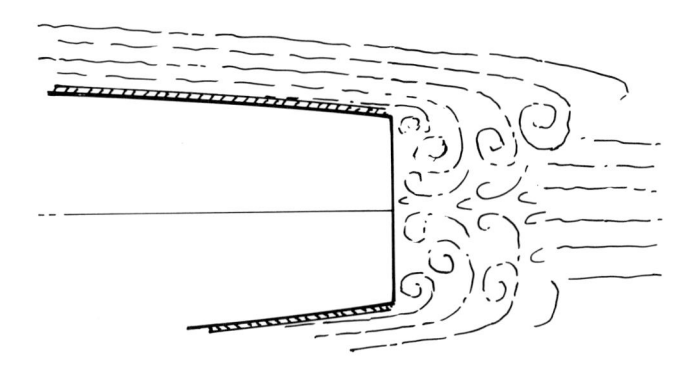

Fig. 13.2 Abrupt changes of curvature such as at bow and stern or transom will produce separation of the water flow and thus, increased drag; a major reason why the lines of a yacht need to be smooth and fair.

In Fig. 13.1 you will note another strange phenomenon: the water is *separating* from the hull due to a shoulder or knuckle in that region. This separation causes *eddies*, which increase the energy loss or drag. The curvature near bow and stern (or a bluff end) are likely areas of separation, as shown in such locations as stern and transom in Fig. 13.2. Now it can be appreciated why the lines of a yacht or boat hull need to be smooth and fair, and free from projections or roughness if it is to perform efficiently.

Another adverse effect on performance is the creation of waves when the yacht is in motion. When the hull drives through the water two pressure points are formed, one at the bow and one at the stern. From these pressure points spreads a definite pattern of waves. Looked at from above, the pattern is a triangular scheme of divergent waves within which are contained transverse waves (see Fig. 13.3). In common with all waves they have a specific length, dependent on the square of the speed at which they are

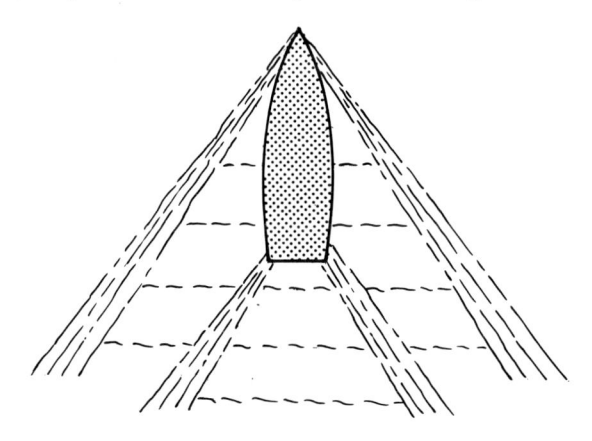

Fig. 13.3 A yacht in motion produces a scheme of triangular, divergent waves within which are contained transverse waves.

As the Waters Slip By

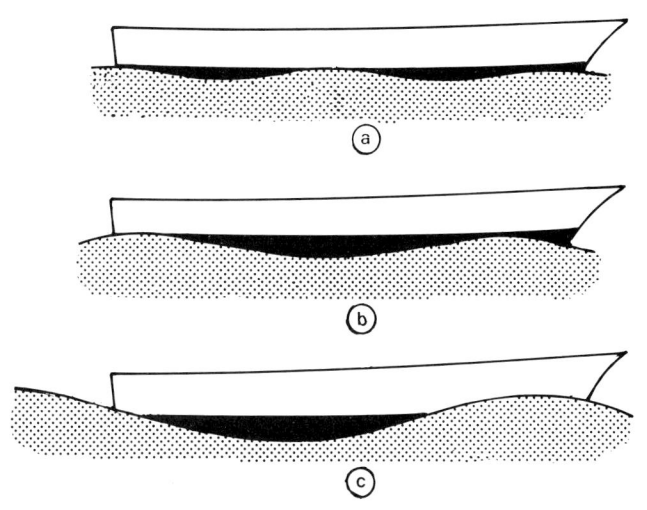

Fig. 13.4 Yacht speed in relation to the wave system is an important element of design. As speed increases so the stern squats in the trough and the boat has to power itself up, or through the wave.

travelling. This speed increases with yacht speed. One point that is important later in the chapter is that the bow wave system starts with a crest, the stern with a trough.

Considering the yacht speed in relation to the wave system, as it increases so the stern squats in the trough and the boat has to power itself up, or through the wave (see Fig. 13.4). From this point on, massive amounts of power are required for a very small increase in speed, unless the hull is of the *planing* type. This critical speed for displacement-type craft is known as the *hump speed*. This will also be discussed later in the chapter.

One other aspect of the wave system created by a hull is very relevant to design. If the crest of the bow system coincides with the trough of the stern system (see Fig. 13.5(a)), they cancel each other out, thus saving energy and improving performance. This highly desirable state is very much a matter of the correct choice of length and speed.

As a yacht hull progresses through water so the wave pattern along its hull will change as its speed increases. This is illustrated in Fig. 13.5(b), which shows a series of *humps* and *hollows* with increased speed. In the diagram, N is the number of waves formed along the speed range and it should be noted that N is an *odd* integer at humps, and *even* at hollows. Where $N = 1$ this is known as the *main hump* because for displacement hulls power is at a maximum for very little speed advantage; where $N = 2$ it is called the *prismatic hump* because its influence is greatly affected by the prismatic coefficient. A naval designer usually wants to be in a hollow region, although other considerations may be overriding in deciding the length of yacht.

The bow and stern pressure systems are about $0.9L$ apart and mathematical considerations of *trochoidal* waves – which closely correspond

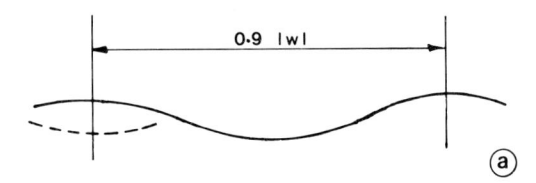

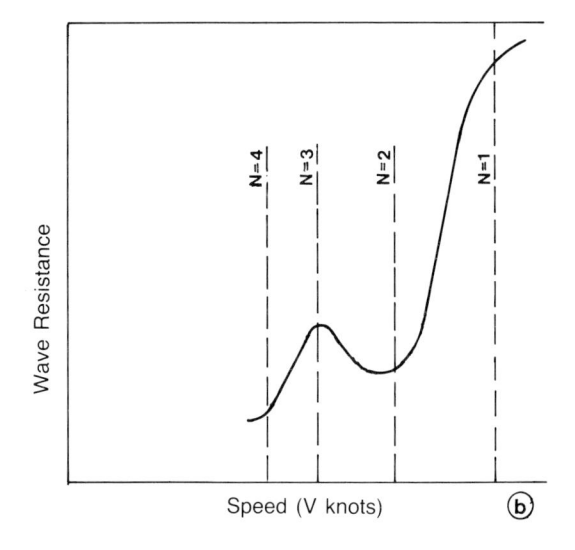

Fig. 13.5 Wave formation due to the motion of a yacht is another important consideration in the efficiency of a yacht hull. The correct choice of length will minimise the adverse drag effects of bow and stern waves formed during the motion of yacht.

to those generated by ships and small craft – provide a formula which gives the condition where the crests or troughs of the bow system coincide with the first trough of stern. In a very simplified form, where V is in knots and L (the LWL) is in feet, this is given by:

$$L = 0.31NV^2$$

If we assume a speed of 6 knots and a value of $N = 2$ (at a hollow), we get

$$L = 0.31 \times 2 \times 6^2 = 22.3 \text{ ft}$$

The hump speed – so critical to design – can easily be found via the speed:length ratio as it occurs at about $V/\sqrt{L} = 1.34$. Referring to Table 13.1, if we take a 25 ft yacht (its square root being 5), then calculate 5 × 1.34, we get 6.7 knots. In fact, efficiency starts falling off rapidly at about 1.0 onwards for our 25-footer, say from 6.75 knots. How far beyond a ratio of 1.34 a displacement boat can drive depends on its shape and power. A thin, sleek forebody with a broad, flattish transom to prevent squatting may reach 1.7, and for a very light displacement boat, about 2.0. But by that time it is on the verge of planing so, if not designed for this, it simply will not go faster.

This creation of waves which causes *wavemaking resistance* is very important for small craft, but there are two other resistances that hold a yacht hull back: *skin* or *friction* resistance caused by the water as it flows past the hull, and *body* resistance due to its shape. The mechanics of friction resistance is described at the beginning of this chapter and this, together

Table 13.1

Water Line Length	Speed/Length Ratio (V/√L)				
	1.0	1.34	1.5	1.7	2.0
	at Speed Kts. of				
16	4.0	5.4	6.0	6.8	8.0
18	4.2	5.7	6.4	7.2	8.5
20	4.5	6.8	7.7	7.7	9.0
22	4.7	6.3	7.1	8.0	9.4
24	4.9	6.6	7.4	8.3	9.8
26	5.1	6.8	7.7	8.7	10.2
28	5.3	7.2	8.0	9.0	10.6
30	5.5	7.3	8.3	9.4	11.0

with the body resistance, is collectively known as *drag*. Drag can be calculated to a large extent, but to obtain the resistance due to wavemaking requires model experiments in a tank. In order to discuss this we need to look into the history, background and techniques of model-yacht and ship testing.

Torquay is a seaside resort on the South coast of England; it also has its share of yachting facilities. Most vistors, or residents, would probably not be aware that it is the place where the world's first ship-model tests were carried out, leading to a scientific foundation upon which the total resistance of hulls could be determined. This enabled power requirements to be obtained. Even more, the man who conducted the tests provided the cornerstone to modern naval architecture.

William Froude was the instigator of the theory and practice of ship-model testing. An eminent engineer who once worked under Brunel and attended the trial voyage of the *Great Eastern*, he was born in 1810 and obtained a first class degree in mathematics. His first claim to marine fame was when he stated – and demonstrated – his famous Law of Comparison which, formally presented, says:

'. . . *that the entire resistance of a ship and similar model are as the cube of their respective dimensions, if their velocities are as the square root of their dimensions.*'

From this statement emerges the speed:length ratio and the idea that by running a model at a comparative speed to that of the prototype ship, the total resistance can be obtained. Using speed:length ratios in the same units, the comparative model speed can be obtained by the following approach:

$$\frac{V}{\sqrt{L}} \text{(ship)} = \frac{v}{\sqrt{l}} \text{(model)}$$

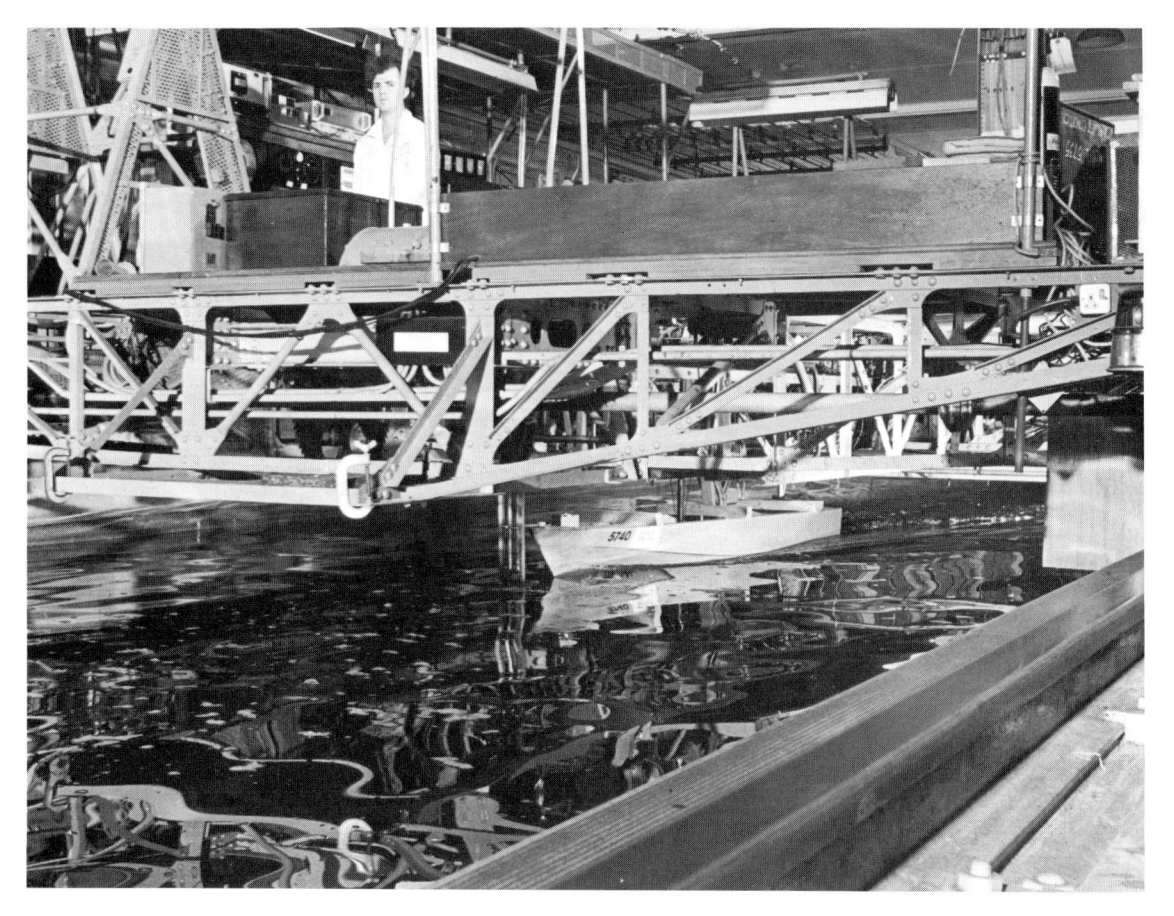

Plate 17 William Froude was the instigator of model testing of ship and yacht hulls. The above tank is one of the model testing facilities of British Maritime Technology, Ltd, one of the major marine research consortia in the world. (*Courtesy British Maritime Technology Ltd.*)

From this we can obtain the model speed, knowing the required yacht or ship speed. As an example, a 25 ft yacht with a speed of 600 ft/min will give the following model speed for a $\frac{1}{5}$ scale model:

$$\frac{600}{\sqrt{25}} = \frac{v}{\sqrt{5}} \quad (\tfrac{1}{5} \text{ scale gives a 5-ft model})$$

$$\frac{600 \times \sqrt{5}}{\sqrt{25}} = v$$

The answer will be about 268 ft/min.

With his son Robert as assistant, Froude worked on a programme of tests for seven years, studying the surface friction of wood planks, finding the variations of resistance with speed, type and length of surface. The values he obtained are still in use to this day. He also demonstrated that the hull resistance could be separated into a *frictional* component and a *residuary resistance* component due to wavemaking and other factors (e.g. eddy drag).

The significance of this work is that, using Froude's values obtained from his tests and calculating the underwater shape of hull, the frictional resistance can be calculated. In a model test the *total* force required to drive the model can be recorded on an instrument called a dynamometer. If the calculated frictional component is subtracted, this leaves the residual resistance. Because Froude's values were for long, thin planks it has been found that between about 4% (fine forms) to 25% (bluff forms) has to be added to the frictional resistance calculated. The wavemaking resistance for a model so obtained is then scaled up for the full-size hull, whose frictional resistance can obviously also be calculated. So the total hull resistance can be estimated. This means the powering requirements of a hull are obtained to a high degree of accuracy, and in the process the hull can be modified for optimum performance.

It is not within the scope of this book to enter into the theory and calculations involved in model-testing, or the numerous research data, graphs and tables that have been collated over the years around the subject. It may be said that in the USA there is a slightly different approach in presenting model data; one eminent US researcher in this field is a man called Taylor, prominent in the study of hump speeds. More theoretical information is given at the end of the book.

Model tests are recommended in the case of expensive yachts or those entering major competitions, but for Mr Average they are hardly necessary – or within his design budget.

14

SHAPING UP TO THE SEAS

Finding the Form Characteristics of Design

Having written a statement of our requirements – the specification – we arrive at a crucial stage in our design. Before putting pencil to paper we must obtain a definite idea of how the germ of our design is to shape up to the seas.

We need to determine the dimensions and general form characteristics of the yacht such as displacement, weight, centres of buoyancy and gravity, if possible, and a rough scheme of hull shape. We also require an estimate of power (if an engine is installed), together with stability aspects, if obtainable.

These considerations will involve some empirical formulae that may make the mathematically faint-hearted run for the first sea novel at hand. Please don't! The formulae should hold no terrors – especially when backed up by simple examples and the mental calculation tools we acquired from Chapter 12.

Early designers had no such calculation aids and relied on intuition and experience to attain their design aims (shipbuilding only became a science in the latter half of the 17th century). Very often, military or trade requirements – or just sheer ignorance of modern naval architecture principles – produced cumbersome vessels that were hydrodynamically inefficient, meaning they were not designed for peak performance in a watery environment. The clipper class of vessels was the first step to such efficiency. The clipper ship era began in the mid-1840s and ended in the United States in 1857. The clipper hulls were those of sea greyhounds compared to the wooden tortoises that preceded them. Nowadays, hull form is no longer subject to random whims but has become part of the art – if not science – of yacht and ship design.

Where does one start in deciding hull shape? Obviously, the basis design is of great help, but if this were the ultimate answer there would be no need to design your own yacht (or any other) because perfect designs would already exist. This is *not* the case – and probably never will be. The perfect design does not exist due to the fact that there is an infinity of combinations to meet individual needs, as well as the service and performance demands to satisfy such needs. So the basis design can only *assist* you in your individual thoughts towards an original design.

In obtaining the form characteristics we will be using the coefficients of form described in Chapter 2. Of course, the coefficients are useless without some guidance on the relevance of the numbers obtained, and how they are to be used. The empirical formulae – another aid to our DIY task – cannot cater for every type of vessel, and so should be used with discretion. If more than one formula exists to determine a certain aspect of design, I will present them all. As the saying goes: two heads – or two answers – are usually better than one. In general, the purpose will be to simplify and streamline the initial design task. So to begin!

Before doing so, let me relate a tale of a young man who had been in the aircraft design industry and was joining (for reasons I forget) a small boatyard where I once worked. He entered the design office fresh-faced, confident and wearing a smart, blue suit. Now, *everything* in aircraft design is spelt out (I know; I once worked for a brief period in the business). There are handbooks that list shapes, sizes and fittings, give information in meticulous detail on the strength and treatment of materials – and even on how to draw the plans. Little room is left for manoeuvre. Although somewhat monotonous for the average designer (who is not forever designing space shuttles or Concorde), he doesn't have to tear his hair out over design frustrations.

I looked on this young man's first efforts in marine design with interest. He laid out his instruments with clinical precision, stacked a mass of useless reference and handbooks on a shelf, then flicked imaginary dust off his drawing stool with a lily-white handkerchief before coming over to me for his first task.

'Go down to the boat,' I instructed him, 'and measure what room there is between the engine and aft bulkhead. I want to see if we can get a pump mounting on the port side.'

It took a while to explain the technical jargon, but when I did he looked at me blankly.

'Can't the drawing tell us that?'

I smiled. 'Theoretically, it should. But workmen don't – or can't – always follow the plans and might have taken a pipe through or built a bit of structure in the way.'

'But that wouldn't *happen* in aircraft design,' he protested.

'Neither do boats fly – or planes float. Now get down there – and don't forget to take a measuring tape.'

He departed, my scornful rebuke ringing in his ears. Two hours later he returned, less confident or button-smart than earlier.

'I've got the measurements,' he wheezed, honest sweat dripping from his brow.

'Good! You can start on the mounting drawing, then.'

Five minutes later he was back.

'Isn't there some sort of manual or book I should work to?' he whimpered.

'No,' I replied, patiently. 'There isn't.'

When he arrived at the office next day his gear was more workmanlike: an old flannel jacket and baggy pants. By the end of the day his drawing instruments were less clinically precise (and fewer in number), while most of the books had disappeared from his shelf. The man was learning to be a yacht and boat designer.

The point of this story is that most of marine design is a very flexible process that involves compromise and continuous adjustment during the preliminary and through to the final stages of creation. There are no absolutely standard procedures – or rigid answers – no perfect solutions, and certainly no final design until the spiral route has been traversed.

So let us now shape up to the seas.

First we need some approximation to the eventual form of hull, based on the form coefficients (you may have to look up Chapter 2 to refresh your mind on their definitions and formulae). The C_b for yachts usually ranges from about 0.45 to 0.55 and has a mean of around 0.50. Racing yachts of very slender form and deep keel may be even lower than this, possibly down to 0.2 in extreme cases. River and estuary cruisers – which come close to being floating, mobile homes – have higher C_b values because of accommodation demands.

The C_p is a measure of the ends of the hull and can be important in powering calculations. Expected values generally exceed 0.55. For sailing yachts with fine-ended hulls the best values seem to lie between 0.52 and 0.54. It is a very important coefficient and Fig. 14.1 shows a curve of C_p against $V/\sqrt{L}$ which is suitable for conventional yacht types.

The C_{wp} is a measure of the *fullness* of waterplane and varies from around 0.70 for fine forms to 0.80 or more for very bluff boats.

The C_m gives an idea of the *fullness* of shape up to the LWL at amidships. Its value can be as low as 0.40 for sailing yachts of very fine form, up to as much as 0.80 or more for river cruisers.

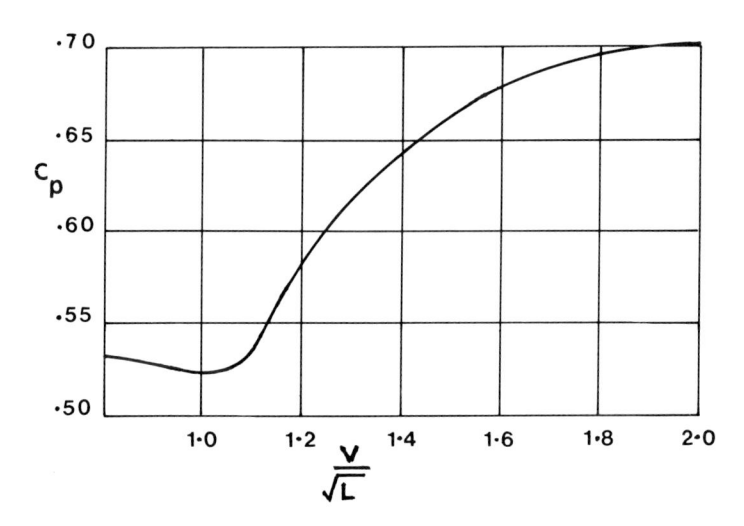

Fig. 14.1 The Prismatic Coefficient (C_p) is important in powering calculations. The above graph shows a plot of C_p against speed-length ratio, suitable for conventional yacht types.

Table 14.1

B = Beam, d = draught, L = L.W.L., ∇ = Volume of displacement Aw = Load waterplane area, C_{np} = waterplane coefficient and C_b = block coefficient. All dimensions are in feet.

Approximate Formulae for dimensions
(a) $B = \sqrt[3]{L^2}$

(b) B (for racing yachts) $= \dfrac{L}{4} + 2.5$

Approximate formulae for displacement and coefficients

Δ (in tons) $= \dfrac{L\,B\,d\,C_b}{35}$ (for salt water)

   divide by 36 for fresh water

$Cwp = \dfrac{2}{3} C_b + \dfrac{1}{3}$

Approximate formulae for Stability (see Ch 21 for application)

Morrish Formula

(c) V.c.b. (below L.W.L) $= \dfrac{1}{3}\left(\dfrac{d}{2} + \dfrac{\nabla}{A_w}\right)$

BM (transverse) $= \dfrac{B^2}{Kd}$

Where K is a constant based on geometrically similar hulls and is usually between 10 and 15

BM (longitudinal) $= \dfrac{3A_w^2 L}{40 B \nabla}$

As mentioned previously, the form coefficients are extremely useful in estimating the limits within which a desired hull should comply and some empirical formulae are presented in Table 14.1 to obtain a rough idea of certain of the coefficients. But they in themselves do not give the yacht dimensions. These have to be decided with the aid of the basis design. There are also empirical formulae that can assist stability decisions and these are also presented in Table 14.1.

One formula (mentioned in Chapter 12) that has been derived from mathematical considerations, and is vital to the relationship between length and speed, is the *Speed:Length Ratio*, $V/\sqrt{L}$. Figure 14.2 shows the relationship between $V/\sqrt{L}$ and hull, plotted to a vertical scale of engine horsepower/boat tonnes. The graph gives some idea of the speed:length regimes for conventional hull forms.

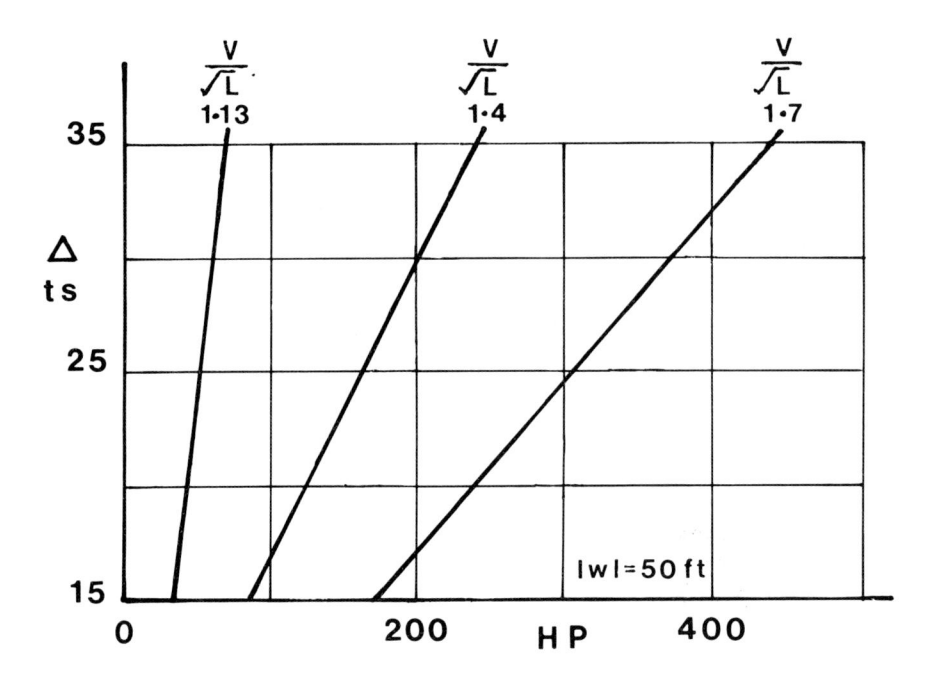

Fig. 14.2 Speed-length ratio is intrinsic to hull type and form. The above graph shows $V/\sqrt{L}$ plotted against engine horsepower/boat tonnes.

Later, we will discuss the relationship between $V/\sqrt{L}$, hull resistance and power, as well as how it was derived, but for the moment we need to consider it in terms of our design length. Before going into the design considerations of $V/\sqrt{L}$ let me repeat a simple example of its calculation, bearing in mind the units have to be knots and feet, as first established by William Froude, the originator of the ratio.

Suppose we have a 25 ft yacht that is expected to achieve a speed of 6 knots, then its $V\sqrt{L}$ value should be $6/\sqrt{25}$. Now if a number, say five, is multiplied by itself it will give its square, or 25. The reverse process is called the *square root* ($\sqrt{\ }$), in our case five. For less simple numbers we would obviously need to use tables provided by most appropriate handbooks and text books or a calculator, which gives an immediate answer when pressing the square root button. So our $V/\sqrt{L}$ will be $6 \div 5 = 1.2$.

Sailing yachts require a hull that is best over a speed:length ratio range between 1.0 and 1.3, though when tacking or in light winds values can be much lower. For motor sailers it should be between about 1.3 to 1.5, depending on length and the compromise between power and speed. For motor yachts at low to moderate speeds it should be between 1.1 to 1.4, depending on hull type, possibly higher for faster craft. In any case, it should never exceed 2.0 for conventional hull forms. Above this we require a planing hull or else the power demands will be impossibly high for very little gain in speed. One formula to obtain speed or length, recommended for yachts, (derived from the $V/\sqrt{L}$ ratio) is given by:

$$V = 1.34\sqrt{L}$$

Before proposing an approach to obtaining the preliminary design characteristics and dimensions, a word should be said regarding the empirical formulae presented in Table 14.1, together with examples, as well as some other design aspects. Formula (a) is an approximate formula I have derived to obtain beam and should be used with a certain discretion, according to whether the design exceeds conventional forms. Formula (b) applies purely to sailing yachts and will be found to produce unexceptional proportions giving good, average ability under all conditions for yachts between 7.5 m and 18.5 m (about 25 ft – 60 ft). In general, the LWL/Beam proportions for sailing yachts range between 2.6 and 3.7. Moorish's Formula (c) will provide a reasonably good estimate of the vbc below LWL but unfortunately, there is no similar one for lcb, yacht and boat types – and their hulls – being so varied in shape. It can be said, though, that most positions of lcb are close to amidships, and with increased speed its optimum position travels aft. At $V/\sqrt{L} = 1.0$ it can well be 1% – 2% of the LWL aft of amidships, which is the usual range for motor yachts. At $V/\sqrt{L} = 4.0$ it can reach as far as 4% aft of amidships. One more dimensional ratio of use as a guide is B/d, which is usually between 2.5 and 3.5 and averages about 3.0 for many boats.

And now to consider the calculation and selection of dimensions and displacement for our design. Obviously, there are a number of approaches to this exercise but one thing is certain – we need to choose some of the design ingredients on the basis of speed, efficiency, space requirements, etc., before we can obtain the others. For instance, a length can be obtained from the speed:length ratio if we have chosen a suitable speed and $V/\sqrt{L}$ number. Or else we can establish length from a preliminary GA sketch, select a suitable $V/\sqrt{L}$ number, and then find speed. For beam we may use a simple rule-of-thumb guide that says that for conventional hulls it should be a value of about LWL ÷ 3. Other approaches may be to scale in proportion to the basis design, use the formulae in Table 14.1, or employ a combination of all the above methods. Because our self-design approach needs to be as straightforward as possible, I suggest the following simple method, with calculations contained in Table 14.2. It requires certain information from the basis design, as presented in the table.

Referring to the table, Step 1 requires that we calculate the $V/\sqrt{L}$ for the basis design and then use this value to obtain our speed in combination with a length obtained from a preliminary GA sketch. It may be profitable to draw the proper GA plan as explained in Chapter 17, up to the point where length is determined. We then make a simple ratio comparison using the basis yacht dimensions (30/9 and 30/3 in our case) to obtain beam and draught, as shown in Step 2. In Step 3 we calculate the C_b for the basis

Table 14.2

Basis Design

Length W.L. = 30'-0" Beam = 9'-0"
Draught = 3'-0" Displt = 9.25 tons
Area (Mid. Sect) = 18 ft^2
Area L.W.L. = 200 ft^2
Speed = 7kts.

New Design

Length W.L = 25'-0"

Step 1 $V/\sqrt{L} = \dfrac{7}{\sqrt{30}} = \dfrac{7}{5.5} = 1.30$

$V = 1.30 \times \sqrt{25} = 6.5$ kts

Step 2 $\dfrac{L}{B} = \dfrac{30}{9} = 3.33$

$B = \dfrac{25}{3.33} = 7.5$ ft

$\dfrac{L}{d} = \dfrac{30}{3} = 10.0$

$d = \dfrac{25}{10} = 2.5$ ft

Step 3 $C_b = \dfrac{9.25 \times 35}{30 \times 9 \times 3} = 0.40$

$\text{Displt} = \dfrac{0.40 \times 25 \times 7.5 \times 2.5}{35}$
$= 5.36$ tons

Step 4 $C_m = \dfrac{18}{9 \times 3} = 0.67$

Area = $0.67 \times 7.5 \times 2.5 = 12.6$ ft^2
Mid Sect.

$C_{wl} = \dfrac{200}{30 \times 9} = 0.74$

Area = $0.74 \times 25 \times 7.5 = 139$ ft^2
LWL

---

design: this is then used to calculate our displacement, using the following formula:

$$\Delta = \dfrac{L \times B \times d \times C_b}{35} \quad (\Delta \text{ is in tons and the dimensions in feet})$$

Step 4 requires that we calculate the form coefficients C_m and C_p using areas obtained from basis design drawings or other sources. For this we need a midship section and LWL contour (from the GA plan, possibly).

All the coefficients can be compared with the information already given – or other sources – to see if they are within acceptable limits or suitable to our requirements. If not, adjustments may be made without too much difficulty. These coefficients are then used to obtain our own design areas.

Shaping up to the Seas

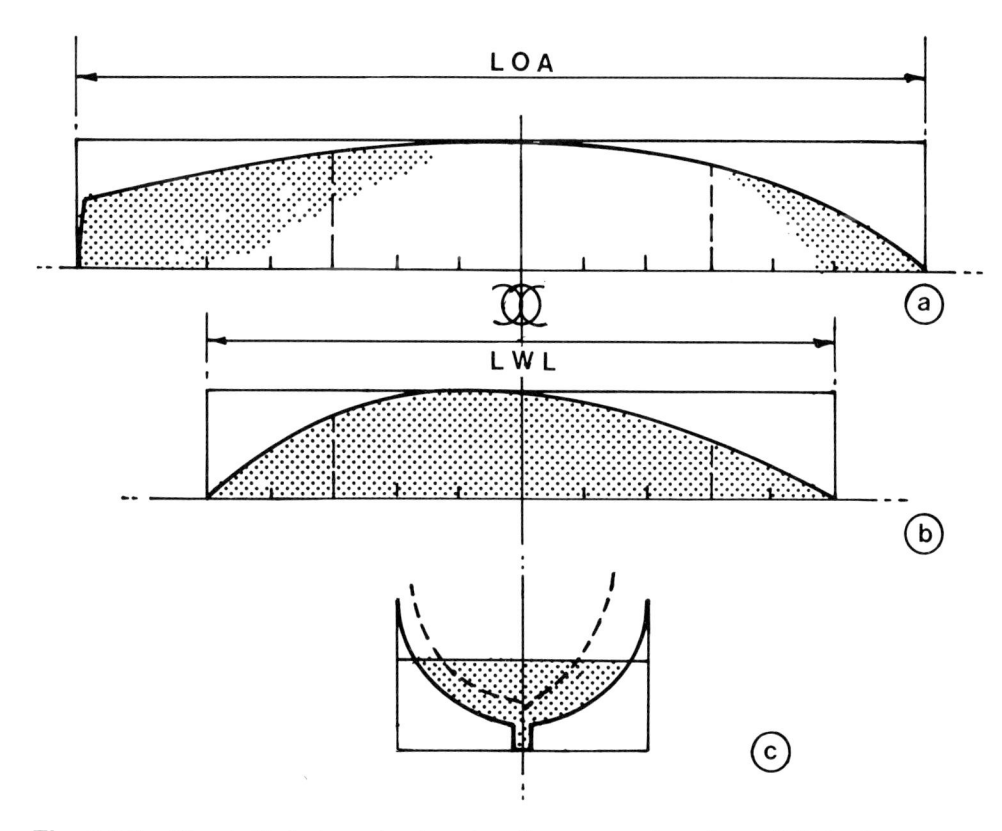

Fig. 14.3 The initial steps in drawing the geometry of our DIY hull design.

So we now have a general idea of the speed and hull form – though only in numerical terms. A detailed graphical picture is provided when drawing the Lines Plan, as explained in Chapter 19, but for the moment we will probably require some idea of what one or two sections and waterlines may look like in order to progress with our GA. Step 4 shows us how to convert the coefficients into areas, so we now make our first excursion into the drawing side of design, as follows (see Fig. 14.3):

(a) Choosing a suitable scale, draw a rectangle using the dimensions of *overall* length and beam. The overall length can easily be obtained using the ratio technique and basis design, as before. Draw another rectangle, only this time using LWL for length. Draw a rectangle using beam and depth as dimensions and drawing a line for draught.

(b) In the small rectangle draw a suitable midship curve that approximates to the required area and shape. It is drawn free-hand at first and the basis design should be of help in obtaining the right shape, which should be continued up to the top of the depth. The area below the draught line (shaded in Fig. 14.3(c)) is then calculated using Simpson's Rule or another suitable method. If the value does not comply with the midship

Design Your Own Yacht

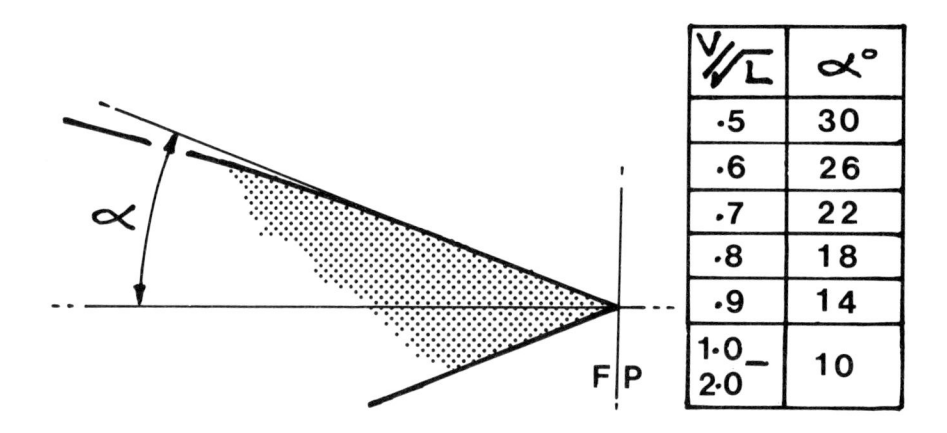

$\frac{V}{\sqrt{L}}$	$\alpha°$
·5	30
·6	26
·7	22
·8	18
·9	14
1·0 – 2·0	10

Fig. 14.4 The angle of entrance of a yacht stem is a main factor in building up pressure at bow. If excessive, hull efficiency can be considerably reduced.

area value calculated from Table 14.2, the curve is judiciously redrawn until, by trial-and-error procedures of calculation and resketching, a shape that suits area and shape requirements is determined.

(c) A similar procedure is carried out for the waterline shape, except that there are restrictions at the bow end which must be complied with for suitable efficiency of form. The *angle of entrance* is the angle formed by the bow at the LWL (see Fig. 14.4), and is a main factor in building up pressure at the bow. If the angle is excessive, hull efficiency can be considerably reduced. The figure shows suitable values of half angle of entrance against speed:length ratio to enable us to draw the bow end of the curve and thus the waterline, as shown shaded in Fig. 14.3(b).

(d) Forward sections for most hulls are usually Vee-shaped and after ones, U-shaped. With this in mind, draw on the midship rectangle one or two sections fore and aft that roughly correspond to the intended shape of hull, again drawing them up to the deck height or top of depth. Make sure you have an idea of where along the length they occur, preferably at a station. Assistance in obtaining their shape may be obtained not only from the basis design, but you will also have a beam dimension from the waterline curve at that location. There are now beam dimensions at certain points to enable drawing a rough plan view of the deck (see Fig. 14.3(a)). With minor adjustments the various curves can be made to correspond with each other, and for further assistance in this procedure read the Lines Plan chapter (Chapter 19).

Shaping up to the Seas

We now have a good idea of hull form; the next initial design requirements are sails and sail power for a pure sailing yacht, or engines and horsepower for a motor yacht. This hull form is now the basis upon which the rest of our design ideas will be formulated. From such an embryonic shell will your own yacht creation emerge – fully-fledged.

15

DRIVING WINDS
Calculation and Design of Sails

Naturally, sailing is very much a matter of practice and experience, but there is an underlying theory. It was Daniel Bernoulli who propounded a theorem in 1738 which not only opened up avenues into the solution of fluid flow problems, but had eventual impact on naval architecture and the revolutionary technology to come – aeronautics. It also put science into sailing.

Within certain limitations, Bernoulli's theorem states that for an incompressible fluid (and water, believe it or not, is nearly incompressible, as you would find out if you landed bellywise on its surface), the sum of the three main energies operating on each fluid particle is a constant. These three energies are (a) *potential* energy (that due to height or position), (b) *kinetic* energy (that due to motion), and (c) *pressure* energy. If one energy is in some way altered then the other two will adjust in rigid obedience to this law of constant balance. As a simple example, if the velocity (speed in a certain direction) of flow is increased (kinetic energy being tied up with velocity), the other two energies will reduce in accordance to maintain the constant.

Now, air at low velocities may be considered as incompressible; it is only when very high speeds are reached (near the speed of sound) that Bernoulli's theorem will not apply. For air (unlike water) we may also ignore the potential energy, its weight being balanced by its buoyancy. So now, if the air velocity is increased round an object the pressure will decrease. Referring

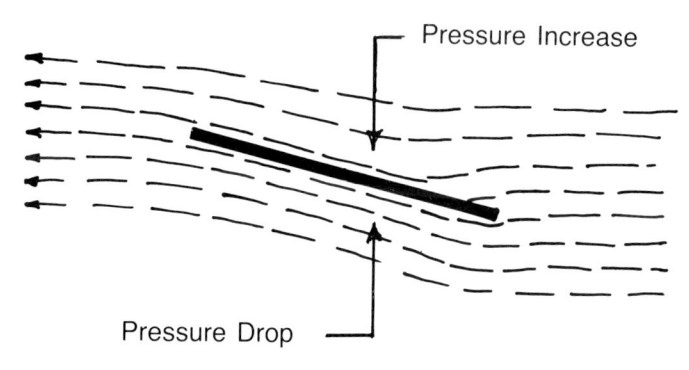

Fig. 15.1 The above diagram shows the pressure difference on a foil such as a sail, when inclined at a certain angle of incidence to the wind.

Driving Winds

to Fig. 15.1, which illustrates a flat foil inclined at a certain angle of incidence to the air flow, the streamline flow at the back of the foil has narrowed due to the increased velocity required to maintain the energy constant. This means there is a pressure drop on this back part of the foil, while on the front side the velocity is slower and thus, pressure is greater. The total effect is to create a *lift* force on the front face. It is this force which makes aeroplanes fly, rudders turn ships, and sails propel yachts when the wind is at an angle.

In fact, there is a total force acting on the foil which can be resolved into a lift force and a drag force. Any force can be broken down into components suitable to a problem. For instance, an object on a slope can have its weight (W) resolved into one component that acts normal or at right angles to the slope (W_n) and one that acts tangentially (W_t) to push it down the slope (see Fig. 15.2(a)). In terms of rudder design it is best to resolve this total force into one that acts normal to the rudder plate and one along the plate; for sail propulsion it is preferable to resolve the force into a *driving component* parallel to the course of the yacht and a *drift* or *leeway component* at right angles to the course.

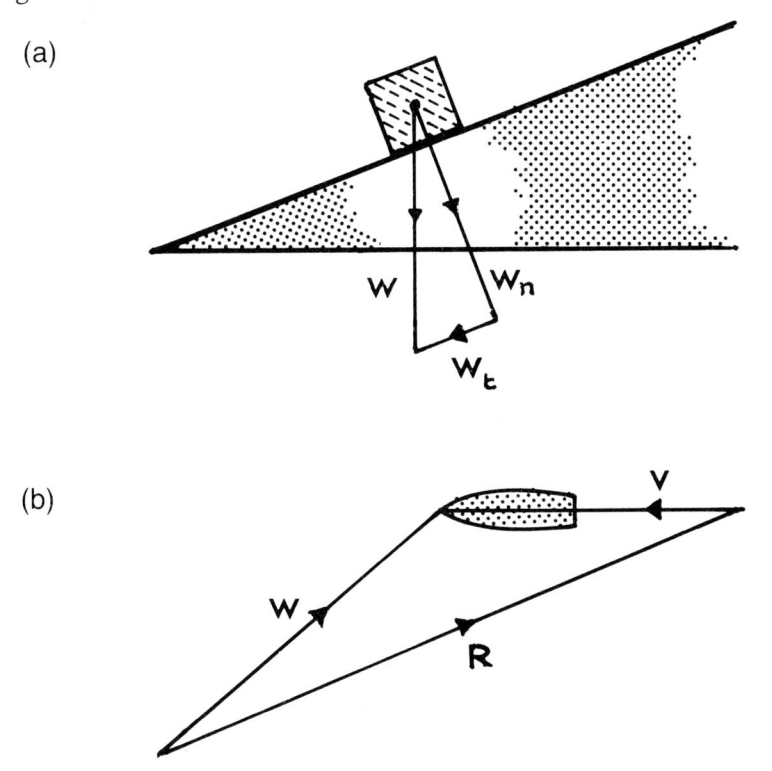

Fig. 15.2 Weights, forces and wind velocities can be resolved into components, as illustrated above. They can be calculated (using trigonometry) or obtained graphically, drawn to a suitable scale.

The force in sailing is, of course, the wind. Unless a wind is blowing from dead ahead or dead astern its effect on the yacht will vary considerably in terms of its propulsive drive. The *true* direction of wind (W) must be resolved into a *relative* wind (R) that is the resultant of the true wind and the speed of the yacht (V), illustrated in Fig. 15.2(b). To avoid the mathematical approach this can be done graphically, drawing the true wind to a suitable scale and resolving the relative wind by drawing the yacht speed on the same scale and completing the triangle.

Now, a sailing yacht or motor sailer is not only subject to the normal forces acting on all surface vessels; it also has air forces acting on the sail and water forces due to its peculiar underwater shape – forces normally negligible for other craft. To achieve the highest efficiency a sail plan has to be shaped to produce the best possible lift-drag ratio from fine angle of attack to the fully stalled state. A formula for the aerodynamic forces is given by:

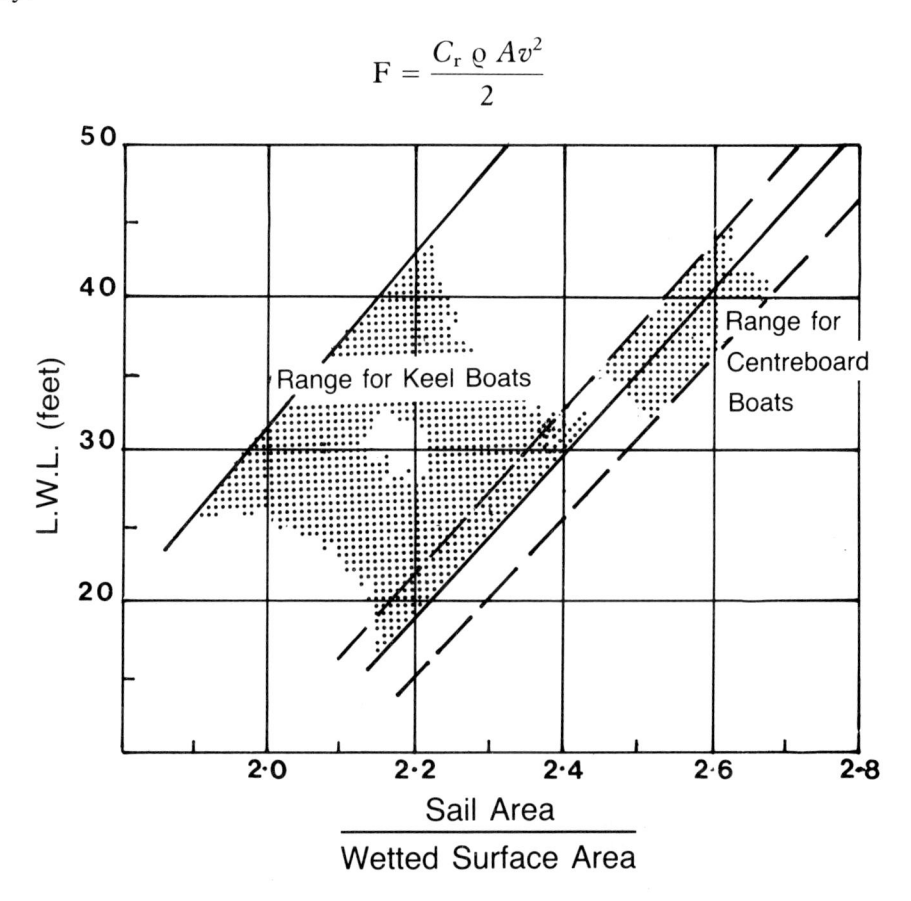

$$F = \frac{C_r \varrho A v^2}{2}$$

Fig. 15.3 One means of obtaining sail area is to use the above graph, calculating wetted surface area of hull by means of Simpson's Rule and methods explained in other chapters.

where F = aerodynamic force and C_r is a coefficient of aerodynamic force governed by the characteristics of the sail plan, A = sail area, v = wind speed and ϱ = density of air. This can also be expressed as:

$$C_r = \frac{F2}{\varrho A v^2}$$

A more theoretical approach is presented in the end chapter, using the above formulae or similar; for our present DIY purposes, and making use of the basis design, a suitable solution is obtained using Fig. 15.3 and calculating the wetted surface area of your hull by principles previously mentioned.

A number of factors enter sail design, including stability and the power to carry sail. One important aspect is balance, which will decide what helm a yacht will take. Balance cannot be predicted with any degree of accuracy, except by tank testing, but two factors are involved in the calculation; the centre of effort (CE) and centre of lateral resistance (CLR). The CE may be considered to act at the centre of area of the sail; for a triangular sail this can be found graphically by drawing a line from each corner (or vertex) of the triangle to a point midway on the line opposite this corner. The intersection of the three lines gives the centre of area. For two sails, and referring to Fig. 15.4, moments may be taken, say, about the mast, and the CE obtained as follows:

$$CE = \frac{(d_1 \times \text{area of mainsail}) - (d_2 \times \text{area of foresail})}{\text{Total sail area}}$$

For CLR, referring to Fig. 15.4, it can be shown that:

$$\text{CLR (about amidships)} = \frac{\text{First moment of immersed longitudinal plane about amidships}}{\text{Area of immersed longitudinal plane}}$$

These values can be calculated using Simpson's First Rule, as described in Chapter 12, only by this time using levers about amidships for the first moment and subtracting the two resultant totals.

If the CE and CLR are not in the same athwartships plane a yawing moment will be produced and needs to be corrected by a rudder moment. If the CE is abaft CLR then the bow will move into the wind and the stern away from it and this can be resisted by giving weather helm, i.e. lee rudder. The opposite will occur for reverse conditions. In this case lee helm, i.e. weather rudder, is required. The distance that CE is forward of CLR is known as 'lead' and its value is purely empirical. It may be unwise to adopt a 'lead' appreciably different from that successful on similar boats.

While this chapter can only sketch over the more complex aspects of sail

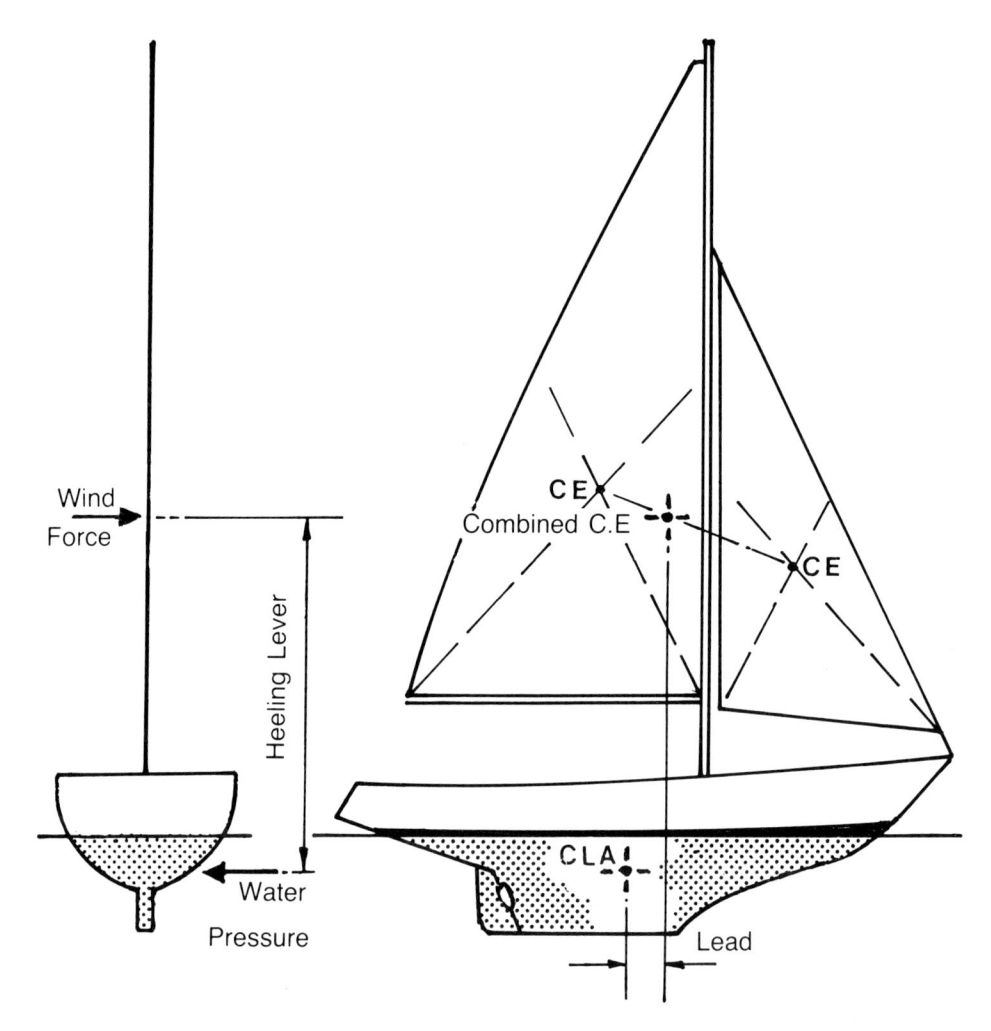

Fig. 15.4 Using the principle of moments – along with the drawing techniques shown above – the Centre of Effort (CE) and Centre of Lateral Area (CLA) can be determined for our self-design.

design, it does provide the basic elements involved and a means of obtaining design requirements. Any further considerations will require more research and specialist information.

With the design parts of the sail now unfurled we can progress under canvas to the other chartered waters of self-design.

16

POWER AND PROP

Estimate of Horsepower and Propeller for Design

The main form of propulsion for most craft – whether motor yachts or large ships – is from a powerplant to a screw propelled via a shaft. The practical elements of this have been discussed in Chapter 7, and we now consider how to estimate the required horsepower for our design if it has an engine, together with the appropriate size of propeller.

The mechanical predecessor to the screw propeller was the paddle wheel. It is claimed that as early as AD 263 there were examples of hand-driven paddle wheels, while from the Middle Ages onwards there are several illustrations of paddle wheel ships. When James Watt (born 1763) practicably applied Thomas Newcomen's atmospheric engine – converting the piston movement to rotary motion – it opened up new vistas for travel, both on land and sea. In 1776 the Marquis Claude de Jouffray built a steamboat in which the engine drove paddle feet. Much later, the *Daniel Drew* of 1860 was able to maintain an average speed of 22 knots.

Then the transition from paddle to prop began to occur. The French scientist Daniel Bernoulli suggested as early as about 1770 a propeller as a means of propulsion and when a propulsive tug-of-war was held by the Royal Navy between *HMS Rattler* (propeller) and *HMS Alecto* (paddle wheel), the result was a resounding victory for the screw-propelled ship.

Among early prop vessels was the *Witch of Stockholm* (1816), and a landmark in this form of propulsion was made by the *Great Eastern* (1859), which was equipped with both a propeller and paddle wheels. Although a commercial failure, its influence in shaping the future of shipbuilding was considerable.

I used to look on marine propellers as strange, corkscrew objects that in some magical fashion were able to push a boat at a rapid pace through water. I then learnt the theory surrounding those blades of curved sculpture. Behind the complex explanations and equations lies one simple fact – they *screw* their way through the medium, providing a forward thrust as they do so.

While there is much calculation, research and development that goes into proper propeller design, together with numerous charts and curves, its selection from our DIY point of view is not so difficult. In fact, even the professional naval architect resorts to the work of propeller experts. Without trying to be funny, there's much room for manoeuvre in the process.

As a case in point, I was not prepared for the following experience with a boat that had just been installed with a new engine at a small yard and marina. During trials the order was given to open throttle for full speed. The speed log dial crept up and up, then stopped – a full half-knot short of estimated maximum speed.

Back at the design office there was consternation and debate, lengthy re-working of the calculations and checking of factors that could pinpoint the problem. At the end a question mark hovered over the diameter of the propeller installed with the engine. A new prop was no answer, considering the time factor involved (the owner wanted to travel to somewhere like the South of France or the Outer Hedrides, I believe), then one old hand came up with an original – and simple – solution.

'Cut a half-inch off each tip,' he suggested, languidly.

You know – that same yacht achieved a half knot *more* than maximum during the next run.

It is not within the scope of this book to cover the pure theory of propeller design, only to give a résumé of the principles involved, some of the nomenclature, and quick, effective ways to achieve the propeller requirements for your design. Basically, the mechanics of prop propulsion are that a column of water (just like a jet) is forced astern to bring about a forward *thrust*. A sternward current is created in the process, known as the *race*. Now, the distance a prop advances in one revolution is called the *pitch*. If it were to advance but deliver *no* thrust, it would travel further than when it did deliver thrust. The difference is commonly called *slip*. Low slip means good efficiency, while high slip provides a large propelling force. These two contradictory requirements must be balanced to the best effect. There are two types of slip, *apparent* slip and *real* slip, but we need not concern ourselves with the subtle distinction between both as it is not necessary to our propeller selection approach.

Because propellers operate in the *wake* of a boat the wake affects their efficiency. This is also taken into account in the propeller calculations and design. The breakdown of these various factors in the propeller operation is illustrated in Fig. 16.1. In terms of practical design, pure theory plays a very small part in the geometric determination of the shape of the blades. Methods of comparison, combined with tank experiments using scale models, provide much better data. Small craft are rarely tank-tested so, taking into account that the power transmitted to a propeller by a small combustion engine is somewhat open to doubt (plus other imponderables such as installation, tuning, etc.), it is absurd to try to aim for a small gain in efficiency and improvement at a high cost in labour and expense. The answer is to use data provided by manufacturers, or easily-read charts that provide the necessary information for a suitable prop.

An understanding of propeller notation is very useful to the selection procedure; if nothing else, it also adds to our total knowledge of yacht design. A right-hand propeller (RH) rotates in a clockwise direction when viewed from abaft, and a left-hand one (LH) in an anti-clockwise direction. Large, single-screw vessels usually have RH screws, and twin screw vessels,

Power and Prop

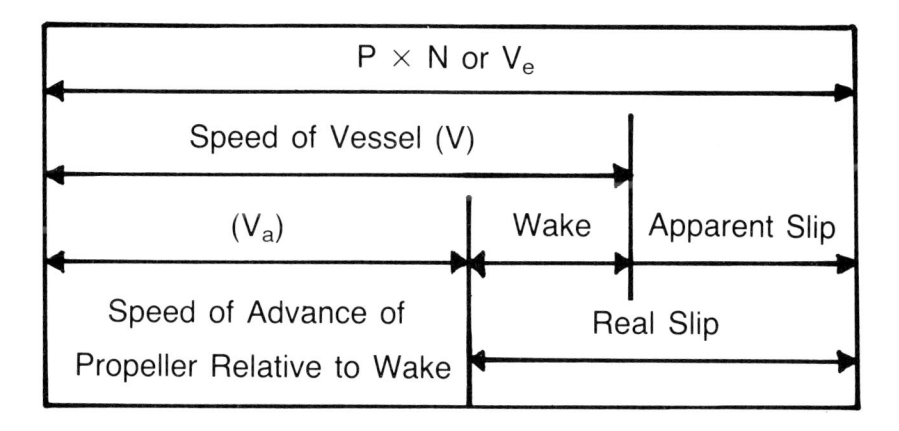

```
|<-------------- P × N or Vₑ -------------->|
|<----------- Speed of Vessel (V) --------->|
|<------- (Vₐ) ------->|< Wake >|<Apparent Slip>|
|<-- Speed of Advance of -->|<---- Real Slip ---->|
|   Propeller Relative to Wake |
```

Fig. 16.1 A number of terms and factors are used in propeller design, as shown above.

one of each. Small craft fitted with twin propellers usually have them *non-handed*, which means they both rotate in the same direction.

A plan of a propeller would usually show the major features and terms as illustrated in Fig. 16.2. *Disc Area Ratio* (DAR) compares the total of actual blade areas to that of the area of a circle of the same diameter, excluding the boss. A propeller is helicoidal in shape and the *geometric* or *face pitch* is the

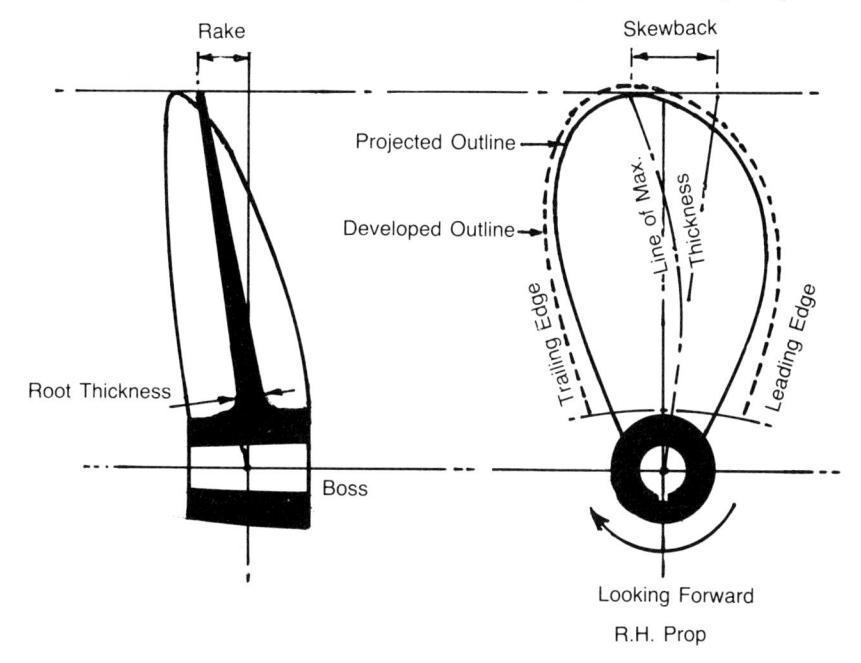

Fig. 16.2 A propeller Plan will have some major features and terms useful to our design knowledge. These are shown in the figure.

105

advance of a blade element in one revolution along a helix of angle the same as that of the blade, as against the *effective* or *analysis* pitch, which is calculated from the revolutions per minute (rpm) for zero thrust. *Rake* is the amount which a blade element leans back from the vertical, while *skew* can be interpreted in a somewhat similar way, except that it is measured along the curvature of a blade. One important ratio is the *pitch ratio*, which is expressed as pitch/diameter. The blade is represented in the figure not only as a *projected* outline (as it would be seen by eye), but also as if it were flattened out on paper to give a *developed* outline.

Before we can make an estimation of propeller dimensions and requirements we have to determine the horsepower that is necessary to drive the yacht at a certain speed. When I was in marine research the bread-and-butter work of the research squad was running models to predict resistance and horsepower of prospective designs. It entailed lengthy preparation, making a rough plastic mould of the model in the model workshop, drawing a Lines Plan to the scale of the model and churning out calculations for the coming tank tests. The Lines Plan was then used to cut the mould to the required shape. This was done by placing the mould in a large bin with twin cutters, fixing the plan on a board outside the bin, then, using a wheel, guiding a pointer round the lines as those terrifying cutters automatically followed the path of the pointer. The process was comparable to driving a car, except that I always grew panicky near the ends of the ship, when I had to rapidly wind the wheel in nervous jerks to keep pace with the motion – and also the pointer on the waterline being cut.

The end result was a close approximation to the final shape, hewn out in steps. The model-makers would then descend on this 'stepped' hull and shave its surface into smoothness. At the tank basin it would then be weighted down to the correct draught and displacement, ready to be clipped under the massive carriage of steel that straddled the long tank of water.

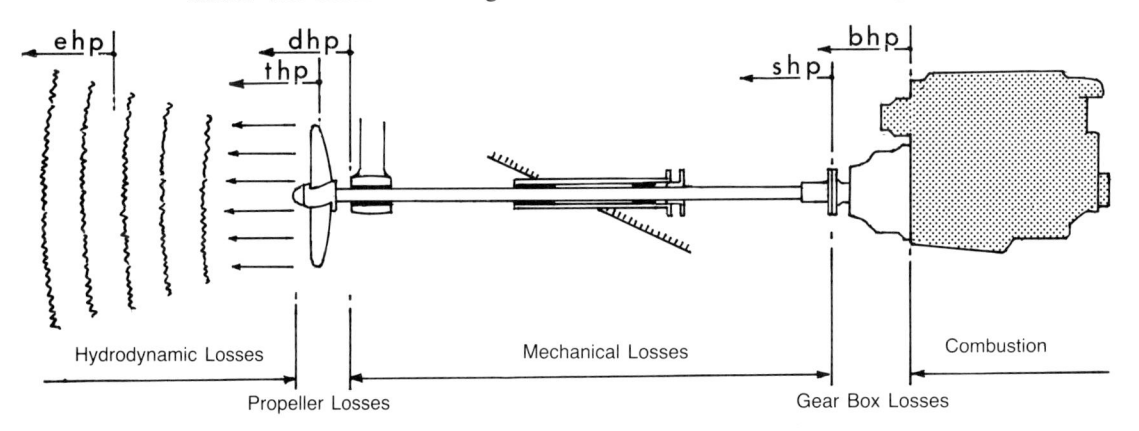

Fig. 16.3 There are a number of horsepower losses from the engine source to the propeller, and the terms used to delineate horsepower along a shaft are shown in the figure.

Running the model at high speeds could be an electrifying experience. With only a few hundred feet of tank to traverse, it was unnerving to be on the carriage and see the end wall approach, wondering whether the carriage driver was awake enough to apply the brakes. One had to be quick to take the values registered by the moving needle connected to the dynamometer and my carriage driver was a placid old boy who read dubious novels during an actual run. How often I remember being delayed in getting a reading, and him calmly pressing the emergency stop button at the last moment. As the carriage came to a shuddering halt (with me perspiring profusely and a storm of waves riding up and down the tank) he would return calmly to his lurid reading matter and suck on his foul-smelling pipe.

The end product of this elaborate experimental procedure was a report and set of calculations that gave power estimates for various speeds – with a little added for various experimental reasons, plus allowance for a margin of safety. While very necessary for large ship designs with overriding commercial considerations, this long-winded approach to finding horsepower is really not applicable to our case.

A number of curves are available to read off horsepower against length, as well as other approaches. One formula recommended is known as the *Admiralty* formula, which relies on similar designs such as our basis design to obtain a constant that will evaluate the shaft horsepower (shp) of our own boat. Before presenting the formula it will be useful to know the terminology connected with horsepower as related to marine engines. Figure 16.3 shows a breakdown of horsepower as delivered by the engine at the combustion stage known as *brake horsepower* (bhp); through the gear box to the shaft as *shaft horsepower* (shp); then delivered to the propeller as *developed horsepower* (dhp); finally to emerge as *thrust horsepower* (thp) and then resolved as *effective horsepower* (ehp) in the water environment, to drive the yacht. The Admiralty formula gives a value for the required shp which is given by:

$$\text{shp} = \frac{\sqrt[3]{\Delta^2} \times V^3}{C}$$

where Δ is in tons, V in knots, and C – a constant – is found from the basis design and then put into our own design figures. Engine manufacturers often do the hard work for their customers in this field, and provide charts and tables to read off horsepower requirements against length or speed:length ratio, for different hull types. Perkins' Engines – a major manufacturer of powerplants for the small craft industry – have kindly allowed their data to be presented in this book, as shown in Table 16.1. The table provides for a range of yachts well within our design purposes. It has been compiled from recorded data by the company and is suitable for most displacement and semi-displacement yacht hulls.

It is impossible to design a yacht which will be equally efficient at all speeds and in this respect the following pointers may be useful. For high speed:length ratios a broad, flat transom is necessary, although it will be less

Table 16.1 Power and speed estimation table displacement and semi-displacement type boats

Type	Displacement						Semi-displacement					
Hull Form	Canoe Stern Narrow Transom						Full Stern Wide Transom Large Under-Water Stern Area					
Waterline length	Disp tons						Speed – Knots					
		5	6	7	8	9	10	11	12	13	14	15	16	17
	0.5	–	–	–	–	–	–	–	15	18	21	25	
	1.0	–	–	–	–	18	21	25	30	35	40	45	
20 feet	1.5	–	–	15	20	25	30	37	45	52	60	70	s.h.p.
	2.0	–	–	20	25	32	40	50	60	70	85	95	
	2.5	–	–	25	32	40	50	62	75	90	105	120	
	3.0	–	15	30	38	50	62	75	90	105	120	145	
	1.5	–	–	–	15	20	25	30	40	45	50	60	
	2.0	–	–	15	20	25	35	45	55	60	70	80	
25 feet	3.0	–	–	20	30	40	50	65	86	90	105	120	s.h.p.
	4.0	–	–	25	40	55	70	90	105	120	140	160	
	5.0	–	15	30	50	70	90	110	130	150	175	200	
	6.0	15	25	40	60	85	105	125	150	180	210	240	
	2.0	–	–	–	15	20	25	35	45	55	65	75	85
	3.0	–	–	–	20	30	40	50	65	75	95	115	135
	4.0	–	–	15	27	40	55	70	85	100	120	140	165
30 feet	5.0	–	–	18	34	50	65	85	105	125	145	170	200 s.h.p.
	6.0	–	15	23	44	60	80	100	125	150	175	200	240
	8.0	–	20	30	55	80	105	130	165	200	235	270	305
	10.0	15	25	40	70	100	140	175	220	260	300	350	400
	3.0	–	–	–	20	25	40	50	60	70	85	100	120
	4.0	–	–	–	23	30	45	60	75	90	105	120	140
	5.0	–	–	15	26	40	55	70	90	110	125	145	170
35 feet	6.0	–	–	17	30	50	65	80	105	135	155	175	205 s.h.p.
	8.0	–	–	20	40	65	85	110	145	175	210	240	270
	10.0	–	–	25	50	80	105	140	180	220	260	300	340
	12.0	–	15	30	60	100	125	170	215	260	305	355	410

Table 16.1 Continued

Type	Displacement					Semi-displacement						
Hull Form	Canoe Stern Narrow Transom					Full Stern Wide Transom Large Under-Water Stern Area						

Waterline length	Disp tons					Speed – Knots						
		5	6	7	8	9	10	11	12	13	14	15	16	17
40 feet	6.0	–	–	15	25	35	50	65	90	105	125	145	165	190
	8.0	–	–	18	30	50	75	95	120	145	170	195	220	250
	10.0	–	–	20	35	65	95	125	150	180	210	245	275	310
	12.0	–	–	23	40	80	115	150	185	220	255	290	330	375
	14.0	–	15	25	47	95	135	175	215	255	300	340	385	440
	16.0	15	18	28	55	105	150	195	240	290	340	390	440	500
	18.0	17	21	32	62	115	165	220	270	325	380	440	500	
	20.0	20	25	36	70	128	185	245	305	365	425	490		
	25.0	25	30	45	90	160	240	315	385	460	530		s.h.p.	
	30.0	30	35	55	110	195	290	370	455	550				

efficient at slower speeds, where a well-rounded hull form – possibly with even a canoe stern – will be most effective. Bear in mind also that, except for drastic changes in midship section area and proportions, alterations will not affect the resistance, and hence horsepower requirements, of the hull. Finally, the shape of sections fore and aft, angle of entrance at the bow, location of lcb and undue projections from the hull, all affect resistance and horsepower and these are discussed in appropriate chapters.

Even supposing a good hull form with minimum resistance and adequate power, a yacht will not achieve optimum performance unless the propeller has the right dimensions. For displacement craft the ideal is to swing as large a prop as there is room for at a speed round about 1,000 rpm, but the propeller must also be chosen so the engine can reach its maximum rated rpm. As many marine propulsion units operate at 3,000 rpm or more, clearly a reduction gearbox is required and ought to be something in the region of 3:1. A Perkins' chart for gearbox selection is shown in Fig. 16.4.

Having found out how large a diameter propeller you can swing, you can play around with the pitch until the engine is producing its optimum revolutions without straining or racing. Large diameter, fine pitched props will have less tendency to slip (and so are more efficient) than small, coarse ones. Planing craft have different criteria and need smaller, higher revving

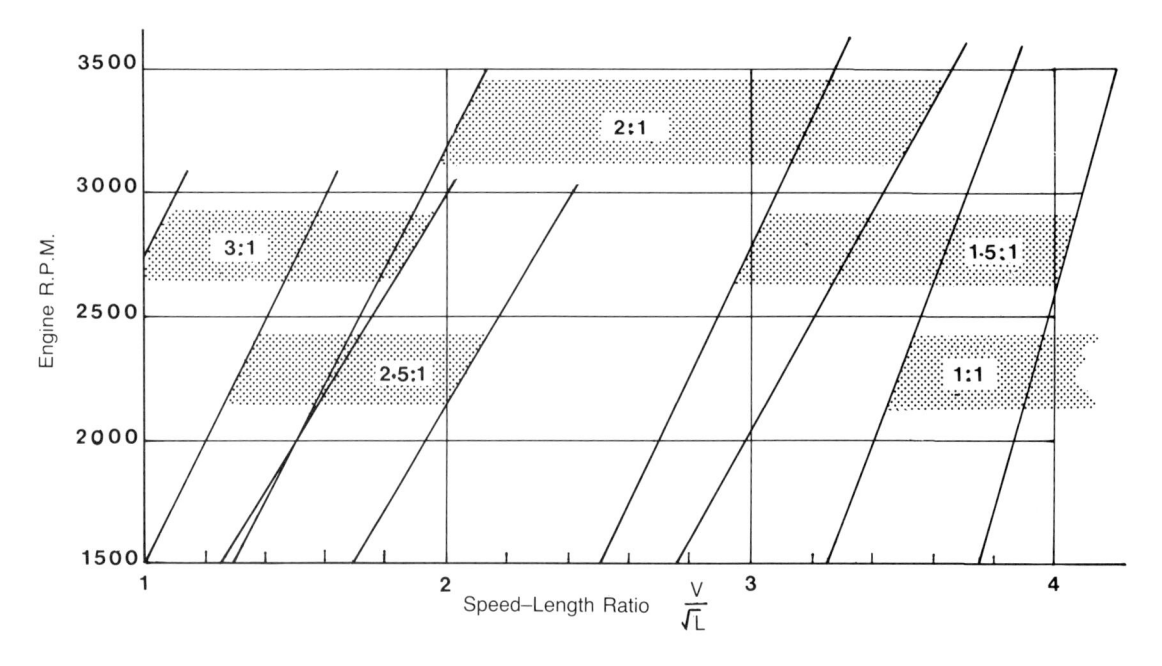

Fig. 16.4 A number of yacht and motor boat engines will require some form of reduction gearbox. The above chart will enable a suitable selection for our self-design.

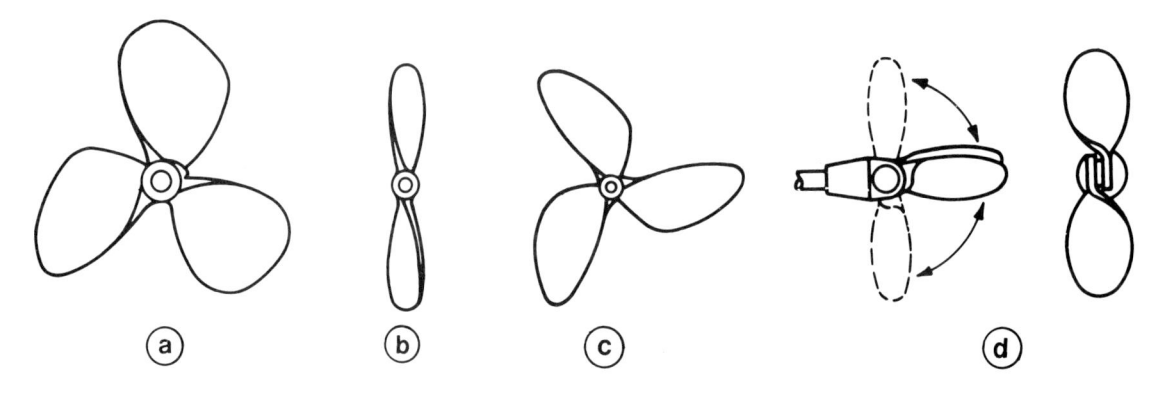

Fig. 16.5 There are various types of propellers employed in the yacht and boat industry. Figure (a) is a high-speed prop while (b) is a screw for auxiliary sailing boats. Figure (c) is a weedless prop and (d) a folding propeller for sailing boats.

Table 16.2

Type of Boat	Wake Factor	
	Single Prop	Twin Props
High speed, racing	0.99	1.0
High speed runabouts	0.98	0.99
Planing cruisers and patrol craft	0.96	0.98
Semi-displacement boats	0.93	0.96
Displacement cruisers	0.88	0.94
Auxiliary Sailing Yachts	0.80	0.90
Fishing and Work boats	0.78	0.89
Trawlers, small tugs	0.75	0.88
Motor barges	0.70	0.85

Va = Boat speed (knots) × Wake factors

and possibly coarser screws; but remember, the use of yacht, its weight and type all play a part in the selection process. Summarising the characteristics of a propeller which contribute to high efficiency, they are: large diameter, low rpm, large pitch, low slip and narrow blades. The fundamentals of propeller selection are diameter and pitch, and we usually have to resort to manufacturers' data or other sources to obtain a suitable prop. Tables 16.2 and 16.3 give propeller data provided by Perkins for pleasure craft, the first enabling you to calculate the required speed of advance for the second. The chart only applies to an 80 b.h.p. engine at 2,500 rpm, but obviously engine manufacturers' data will cover all normal horsepower requirements. As for propeller type, Fig. 16.5 illustrates various types of screws that are employed in the boating industry, especially for pleasure craft.

Small craft with propellers delivering high thrust at high rpm suffer a phenomenon known as *cavitation*, which occurs when the flow of water past the blade surfaces ceases to follow the blades. With this breakdown of flow voids are formed which erode the surfaces, thus affecting efficiency. Cavitation should be avoided at all costs, but it is not possible to enter into this subject in depth here, although it may be said that super-cavitating propellers are specially designed for high-speed craft and those operating in cavitation conditions.

The establishment of power and prop now enables us to enter the depths of self-design, the detailed plans and calculations which give it form, strength and stability.

So we now progress to other elements of our yacht design beginning with the Preliminary General Arrangement, which will be the first paper concept of our design ideas.

Table 16.3 Typical Propeller selection charts

4.236M   60 kW (80 bhp) at 2500 rev/min

Pleasure craft
Full accommodation type offshore cruiser (with planing and semi-planing hulls)
Motor sailers
Light displacement cruisers

Speed of advance (Va) knots	Boat speed knots One engine Wf 0.93	Boat speed knots Two engines Wf 0.96	Propeller data Direct Drive	1.2:1	1.5:1	1.91:1	2:1	2.1:1
6								
7	7.5	7.3				22.5 × 12.5 0.6	22.5 × 13.0 0.6	22.5 × 13.5 0.6
8	8.6	8.3				22.0 × 13.0 0.55	22.0 × 13.5 0.55	22.0 × 14.5 0.55
9	9.7	9.4	22.0 × 13.5	22.0 × 14.0	22.0 × 15.0	0.5	0.5	0.5
10	10.8	10.4		19.0 × 11.5 0.55	22.0 × 14.0 0.45	22.0 × 15.0 0.45	22.0 × 16.0 0.45	
11	11.8	11.5	15.0 × 9.0 0.65	16.5 × 10.0 0.65	19.0 × 12.0 0.5	22.0 × 14.5 0.45	22.0 × 15.5 0.45	22.0 × 16.5 0.45
12	12.9	12.5	15.0 × 9.0 0.6	16.0 × 10.5 0.6	19.0 × 12.5 0.5	21.5 × 15.5 0.4	21.5 × 16.5 0.4	21.5 × 17.5 0.4
13	14.0	13.5	15.0 × 9.5 0.55	16.0 × 11.0 0.6	18.5 × 14.0 0.5	21.5 × 16.5 0.4	21.5 × 17.5 0.4	21.5 × 18.5 0.4
14	15.1	14.6	15.0 × 9.5 0.5	15.5 × 12.0 0.55	18.5 × 14.0 0.45	21.0 × 17.5 0.4	21.0 × 18.5 0.4	21.0 × 19.5 0.4
15	16.1	15.6	15.0 × 10.0 0.5	15.5 × 12.0 0.5	18.0 × 14.5 0.45	20.5 × 18.0 0.4	20.5 × 19.5 0.4	20.5 × 21.0 0.4
16	17.2	16.7	14.5 × 10.5 0.5	15.0 × 13.0 0.5	17.5 × 15.0 0.4	20.0 × 19.0 0.4	20.0 × 20.5 0.4	20.0 × 22.0 0.4

Table 16.3 Continued

4.236M 60 kW (80 bhp) at 2500 rev/min

Pleasure craft

Full accommodation type offshore cruiser (with planing and semi-planing hulls)
Motor sailers
Light displacement cruisers

Speed of advance (Va) knots	Boat speed knots One engine Wf 0.93	Boat speed knots Two engines Wf 0.96	Propeller data Direct Drive	1.2:1	1.5:1	1.91:1	2:1	2.1:1
17	18.3	17.7	14.5 × 11.0 0.45	15.0 × 13.5 0.5	17.5 × 16.0 0.4	20.0 × 20.0 0.4	20.0 × 21.0 0.4	20.0 × 22.5 0.4
18	19.4	18.8	14.5 × 11.5 0.45	14.5 × 14.0 0.45	17.0 × 17.5 0.4	19.5 × 20.5 0.4	19.5 × 22.0 0.4	19.5 × 24.0 0.4
19	20.4	19.8	14.0 × 12.0 0.45	14.5 × 14.5 0.45	17.0 × 18.0 0.4			
20	21.5	20.8	14.0 × 12.5 0.4	14.0 × 15.5 0.4	16.5 × 19.0 0.4			
21	22.6	21.9	14.0 × 12.5 0.4					
22	23.7	22.9	14.0 × 13.0					
23	24.7	24.0	13.5 × 14.0 0.4					

17

OF SHOES AND SHIPS AND SEALING WAX
Drawing the General Arrangement for the Design

>'The time has come,' the Walrus said,
>   'To talk of many things:
> Of shoes – and ships – and sealing wax –
>   Of cabbages – and kings. . . .'

Drawing a General Arrangement plan for any vessel – whether yacht or supertanker – requires the gluing together of many bits and pieces in terms of knowledge, skill and ingenuity; the juggling of space requirements with form restrictions or compromising between speed and safety, cost and comfort. As with the *Walrus and the Carpenter*, the time has now come to talk of many things about yacht design . . .

A good yacht design should be seaworthy, easy to handle, economically fast for its type, and structurally safe and strong. But it also has to carry crew and passengers to a required standard of comfort, as well as equipment and outfit related to its purpose and service. One last point: an artistically pleasing design can but add to its value – if only in aesthetic coinage.

You may well consider that so many divergent – and possibly conflicting – factors cannot by any stretch of the imagination be set out in one go. You are perfectly right. Let me relate a story.

The Chief whom I served under during my training one day started me on my first serious plan – drawing an As-Fitted General Arrangement. This drawing is a graphical document of the craft as built, and my task was to record on blue linen a research trawler that was near completion. It all seemed so simple, making a blueprint copy of an existing vessel.

Diligently, I laid out the linen and let it stand a few hours to allow for shrinkage or expansion, then powdered it in preparation for the ink lines to be drawn. Armed with measuring tape, a preliminary GA plan and note-pad, I vigorously strode out toward the newly-built ship, ready to gather the required measurements and information and transfer it onto the linen in two – possibly three – days at most.

One week later, the Chief paused at my board.

'Hmmm! I see you have drawn the outlines and put in some details,' he grunted. 'Don't forget, the vessel's due to leave in three weeks' time. You *did* say a couple of days was enough for such a copy job, didn't you?'

My reply was not humble – it was positively abject. That week had been a back-breaking grind of crawling down manholes and climbing up ladders,

Of Shoes and Ships and Sealing Wax

Plate 18 Early sailing ships and craft were overcrowded, top-heavy and potentially dangerous in freak conditions, as typified by the 91-gun carrick, *Mary Rose*, shown above, which capsized in the Solent over 400 years ago. (*Courtesy Masters and Fellows, Magdalene College, Cambridge.*)

taking fiddling measurements and scribbling numerous notes. I had never seen so many oddball items of equipment – in the most unexpected places – in all my life. Neither had I so often bumped my head against wicked projections or scraped my shins on pieces of steel that maliciously jutted out. Ventilation and exhaust ducting seemed to coil hellishly into unreachable dark depths; innocent light boxes or gate valves were located in treacherously awkward places.

I was still measuring and crawling (mentally, as well as physically) when the trawler went on trials two weeks later. That one, simple GA had become in my mind a monstrous spider who spun a web of infinite lines and made cleats and brackets spring up everywhere to ruin one's day, or created unsuspected piping that twisted one's patience to shreds. When I finally completed the drawing I felt as if I had *built* the boat as well.

If it's so difficult to draw a General Arrangement for a craft that exists, how much more so for a mere concept or idea? Well, have no fear! If you have paid some attention to the previous chapters, and follow the procedures of this one, you will find it a relatively easy task. Naval architects of the past had to draw some horrendous plans to describe their wooden designs – and they managed them with skill and dexterity.

These plans – which still exist in museums and archives – reveal that early sailing ships and craft were overcrowded, often top-heavy and potentially dangerous in freak conditions. Take the case of the 700-ton, 91-gun carrick *Mary Rose*. This vessel was salvaged from the depths of the English Channel (in the Solent) over 400 years after capsizing. In calm weather – and under the eyes of Henry VIII and his fleet – it turned over in 1545 due to inherently unstable characteristics, combined with freak overloading. The vessel was a converted merchantman.

The wooden galleons of those days were arranged to take temporary castles fore and aft during times of war. These castles later became a permanent part of the structure to give them a cumbersome but fortified appearance in profile. By present standards they lacked the sleek appearance of modern sailers, but remember, their requirements were quite different and the master shipwrights' knowledge of naval architecture quite limited.

Not only did these ships lack in form and appearance, their arrangement and layout also left a lot to be desired. For instance, the lower deck of a 74-gun vessel was both the sleeping deck and mess for the crew, who slung their hammocks and ate their barely palatable meals in semi-darkness for days on end, because in even moderate seas it was necessary to close the port lids to stop the sea water coming in. A dank and noisome place, in hot climes the crew's discomfort was aggravated by the seams of the deck above opening to allow water to drip down on hammocks and sleeping bodies.

We would certainly not wish our design to have such flaws, and the GA drawing ensures that there is comfort, safety and the proper distribution of outfit and equipment, as well as a pleasing appearance. Drawing the plan requires the gluing together of pieces of knowledge and skill, as well as ingenious manipulation of space allowance within the form restrictions. So we now talk of yacht design in terms of shoes and ships and sealing wax – and how it's all put together.

Assuming we have determined the length, beam, depth, and some idea of draught and displacement – as well as bow and stern contours – we now need to draw the Elevation, Plan View and three or four typical sections through the yacht. A 10-metre (33-ft) yacht – using a scale of $1/10$ (or, say, 1" to the foot for Imperial units) – requires a sheet of paper about $1\frac{1}{2}$ metres long (roughly 5 feet), which will also cater for the sections. Other yacht sizes may be in proportion, or drawn to suitably different scales. If desired, the sections can be drawn on a separate sheet.

If a Lines Plan exists we can transfer an outline of the Elevation or Profile and Plan of the deck of our design (maybe even trace it), together with the appropriate datum lines to be discussed next. The two views should be located on the extreme left of the paper, the whole of the right-hand side being used for the sections. If there is no Lines Plan guidance, we need to draw the essential datum lines from which all measurements and other lines flow. These are the baseline – situated about midway on the paper – for the

Elevation, and a centreline – near the bottom – for the deck Plan View, as well as two long vertical lines to represent the perpendiculars. Remember, their distance apart (to scale) is the waterline length, or length between perpendiculars (LBP), whichever is applicable. Another vertical is struck exactly half-way between to represent amidships (⊗). From the Elevation baseline we measure up the scaled draught dimension and draw a parallel line to represent the floating waterline (draught) of the yacht. Our GA drawing task has now begun with the delineation of these very important datum lines.

Our next step is to measure and mark the frame spacing between the perpendiculars – the location of frames or ribs of the hull. This spacing ranges for most yachts between 20 and 45 centimetres (about 8 to 18 inches), depending on size, and an approximation can be obtained from the basis design or else estimated from Classification Society rules. The frames should be equally spaced, with closer spacing at bow to withstand the greater loads. Bear in mind that the *stations* (previously mentioned) used for the Lines Plan and calculation purposes should be located on frame positions for a variety of practical reasons. There are ten equal spacings for the eleven stations so the spacing needs to correspond to these, with minor adjustment possibly being made to the waterline length to obtain suitably simple round numbers for the frame spacing. After all, we don't want to work with numbers such as 25.2 cm or 12¼" when a few centimetres' or inches' difference to the length can round them off to simple figures which are, therefore, easy to handle.

The frame spacing is indicated on the baseline and centreline by small pencil ticks and we then progress to locating the bulkheads and flats. In the strange language of yacht design and naval architecture, the *floors* of a vessel are those parts of frames at the bottom of the hull, while *flats* are just like the floors of a house – what you walk on. In large ships they are called lower decks. But our first job is to locate the main bulkheads.

The main bulkheads of any vessel are the main strength members (athwartships) and are always located in way of frames, replacing them. They are spaced according to accommodation, engine (if any) and other requirements. Some have to be watertight in accordance with safety against flooding etc. The basis design may be helpful in this exercise and starting from the bow, we have the fore peak bulkhead which is usually a few frame spaces aft of the fore perpendicular, say two or three frame spaces, depending on length. In an ocean-going environment this bulkhead will almost certainly need to be watertight in case the bow is breached, but for quiet waters I have seen it pushed further back to make a cramped sleeping space. The fore peak space can be used as an anchor chain locker for large yachts (a table is given in Chapter 24 to calculate its size), or merely as a stores locker with access through the bulkhead,· for such craft as pleasure cruisers on rivers and canals.

The accommodation on both sailing and motor yachts is usually situated at about amidships and towards the bow. The cabin entrance bulkhead is usually the one right aft and if there is only one cabin, the fore peak bulkhead will form the other end of the accommodation. Larger quarters will have intervening bulkheads to divide the accommodation spaces, all located on the frame spacing. There may also be small partition bulkheads – on frame spacing, if possible – to form toilet and galley compartments, etc.

To make a sensible spacing for such bulkheads we need to consider the length of berths – about 180 to 195 cm, or roughly 6' to 6'-6" – the size of clothes lockers (45 to 60 cm or about 1'-6" to 2'), and any other accommodation requirements affecting its length. Once this overall accommodation length has been determined we can then locate the bulkheads on frame spaces that incorporate this length.

If the yacht design has an inboard engine we usually need to encase it in an engine compartment – which means some form of athwartships, and possibly longitudinal, bulkheads. Most inboard engines are located as far aft as possible and lie on a downward sloping slant toward the stern. This sloping line – which is drawn through the line of shafting – is governed by the engine support structure and the clearance of propeller at the stern, among other factors. To position, we can make rectangular cut-outs of the overall dimensions of our proposed engine in the three views and place them in the proposed engine location. Remember, for the Plan and Section at this location we need a rough idea of the hull shape in way of the engine. The Elevation rectangle cut-out – placed in that view – will enable us to place bulkheads either side on frame spaces, allowing enough room for working in the compartment, access to any glands and piping, and so on.

Plate 19 Cardboard cut-outs can be used to locate the engine and other items of equipment on the GA Plan in order to determine their best location and clearances.

The masts of sailing yachts will almost certainly be placed over strength members such as bulkheads, but obviously their location depends on sailing and rig requirements. Finally, we have the aft peak bulkhead (except for very small sailing yachts and dinghies) which is spaced similarly to the fore peak version. It is sometimes also watertight and its space used as a stores locker, if not for steering and rudder gear. Other bulkheads would include those for fuel and water tanks and longitudinal partitions, where rational to the design.

Next we come to the flats. The height of the accommodation flat depends on headroom, as shown on Fig. 17.1, in the process allowing enough space underneath for reasonable depths of floors. The criterion here is that there is enough meat of floor material at the shallowest end. The measurement for height of flat is taken from the baseline and always drawn parallel to it. Engine compartments – if large enough – will also have flats around, or underneath, the engine. These are again dependent on height of floor and engine support structure, but not to such a critical extent. Other flats will be located in wells, wheelhouses and bridges, if these form part of the design. Again, headroom and structural considerations enter the problem.

One last point before we move on to other aspects of the GA drawing: the location of frames and bulkheads at their spacing is to the *after* face of the structure, forward of amidships and the *forward* face after the midships, while the height of flats from the baseline is to their *underside*. As many flats are built in wood, the thickness of the wood needs to be taken into account when considering headroom.

At this stage – sparse though it may appear on the drawing – we have

Fig. 17.1 The height of the accommodation flats depends on headroom, as illustrated above.

delineated the major space requirements on our GA, which should now be of the form shown in the pictorial illustration, Fig. 17.2. Notice that the baseline and centrelines are drawn like this: ⊥⊥⊥⊥; while the bulkheads and flats appear as thick dotted lines in the Elevation and Plan views. This is because they are assumed *behind* the skin of the hull and deck and such lines – in standard drawing practice – are always shown dotted. Views showing the interior arrangement will also need to be drawn on another sheet of paper, but the present drawing is concerned with the exterior layout.

I recall looking at my own GA efforts at this point and thinking how meagre were the results – just outlines circumscribing a few dotted lines. There seemed an endless amount of detail still to be drawn, much of it of a seemingly vague and amorphous nature.

'Don't forget to show the steering and ventilation,' said the Chief Designer when he studied my bare configuration. 'And the rigging and deck equipment, as well.'

I nodded dumbly. The design mountain grew ever more insurmountable.

But with the aid of our basis design, the task is really not as bad as it seems. While it is not possible to write in depth about the various detailed

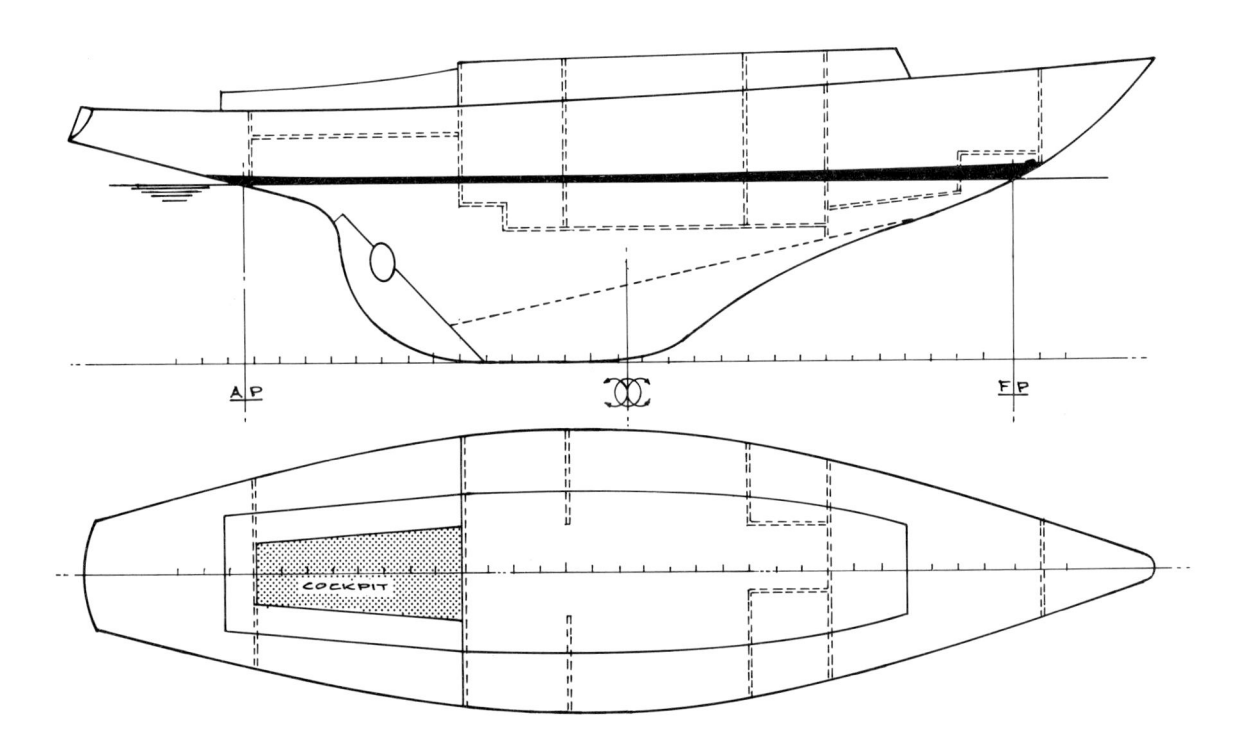

Fig. 17.2 The initial steps in drawing the GA of our self-design is to delineate the major space requirements, frame spacing, bulkheads, etc., as illustrated here.

Of Shoes and Ships and Sealing Wax

considerations within the GA design, I will point out the major considerations, with some guidance in their drawing onto the plan. You can always look up appropriate chapters for further information. The rest is up to you.

The concrete elements in a GA design require filling the various compartments and space requirements with outfit and hardware. This means drawing Elevations, Plans and sections of the interior views in way of accommodation spaces, working areas, engine compartments, bridge and wheelhouse areas, if any. This means we have to draw the contours of the appropriate space at both the top and bottom levels. The previous outlines will partly serve for the hull interior, but contours must also be drawn at the level of flats or deck away from the centreline. Without an existing Lines Plan this obviously means a certain amount of guesswork, but the basis design should be of great assistance in this as its hull form should not deviate too greatly from your own. Allowance must then be made for cabin linings etc., and again, the basis design should solve this problem. Views above deck are much more simple and should offer no great difficulty. Our standing at this stage is shown in Fig. 17.3. Notice the full lines for bulkheads, flats and decks, linings, etc. that can be seen, with dotted lines for those that are covered. Flats can also extend just to the edge of berths in accommodation spaces (see Fig. 17.4).

We now have an approximation to the *true* working space within compartments and need to fill them with the appropriate outfit and equipment. Accommodation spaces will have berths or bunks, lockers, drawers, tables and so on, while the galley – if any – will have a range or stove, shelves and storage space. The toilet compartment will be fitted with a suitable toilet, washbasin, bath (if large enough), and flushing and water facilities. These last two compartments will need to be adequately ventilated.

Regarding accommodation spaces, a length for berths in the range of 180 to 195 cm (about 6' to 6'-6") has already been recommended for comfort, while width should be in the region of 75 cm (about 2'-6"), although a

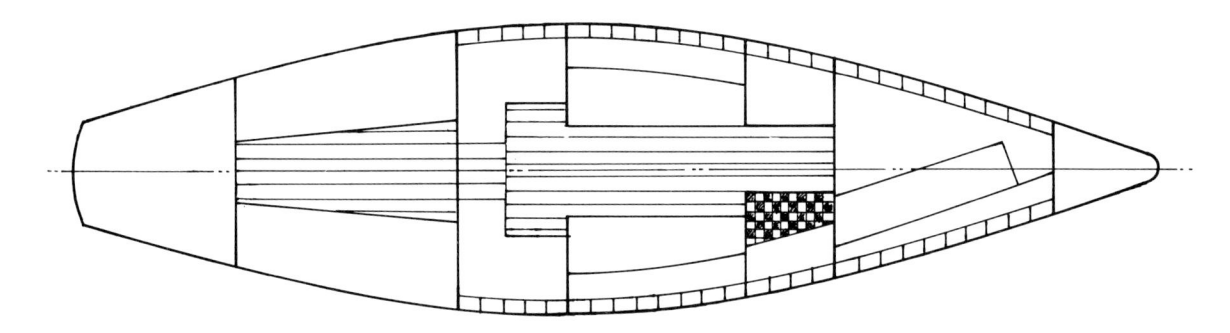

Fig. 17.3 A further stage in the GA Plan is to show the cabin linings, plan views at deck and flats, and other important boundaries of our final space availabilities.

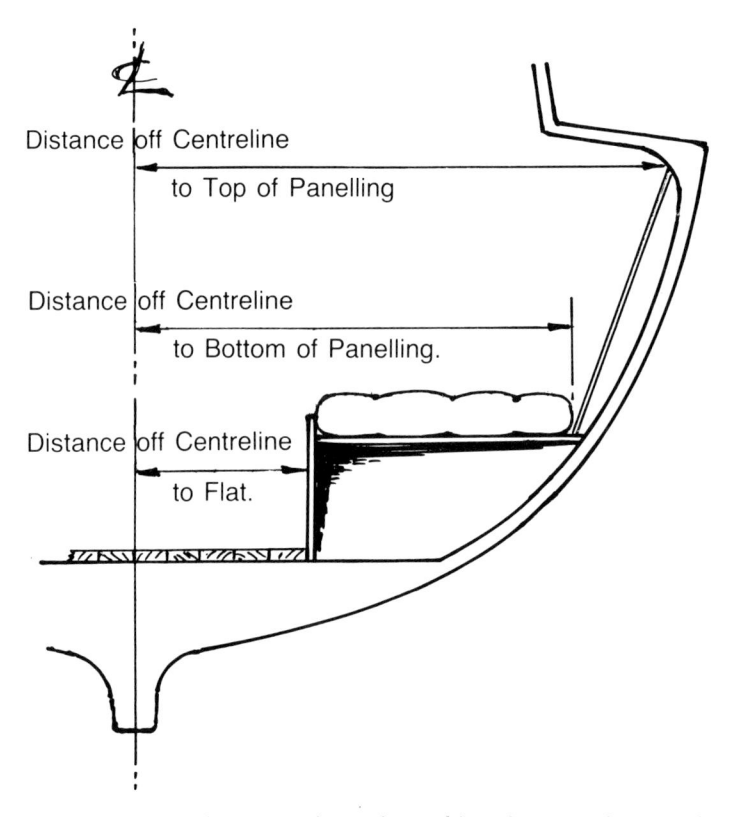

Fig. 17.4 Flats can terminate at the edge of berths, as shown above, instead of projecting to the hull side.

certain amount may be 'pinched' if space is tight. The height between 'stacked' berths depends on drawer space under berths and height of cabin, remembering to leave enough clearance for suddenly awakened bodies not to crack their heads against upper bunk or cabin top.

The accommodation requirements of yachts and boats always present space problems: how can a designated number of people fit into a confined space, and yet achieve some sort of domestic comfort? There is a lot of room for ingenuity here – it does not require high technology. Major considerations are headroom, daylight lighting and floor space, which will influence the cabin superstructure. Space-saving devices such as lower and upper berths and folding tables (the space under lower berths can be made into lockers or drawers) can be used. Berths can be folded up, while lower ones can be turned into settees during the day. A layout of a typical accommodation arrangement and such devices is shown in Fig. 17.5.

Galley facilities depend on the size of your design, requiring reference to brochures and catalogues for a suitable stove or range. Storage space is usually tight in this compartment, so a certain amount of innovation is required in your layout. Similar reference to trade literature is needed to

Of Shoes and Ships and Sealing Wax

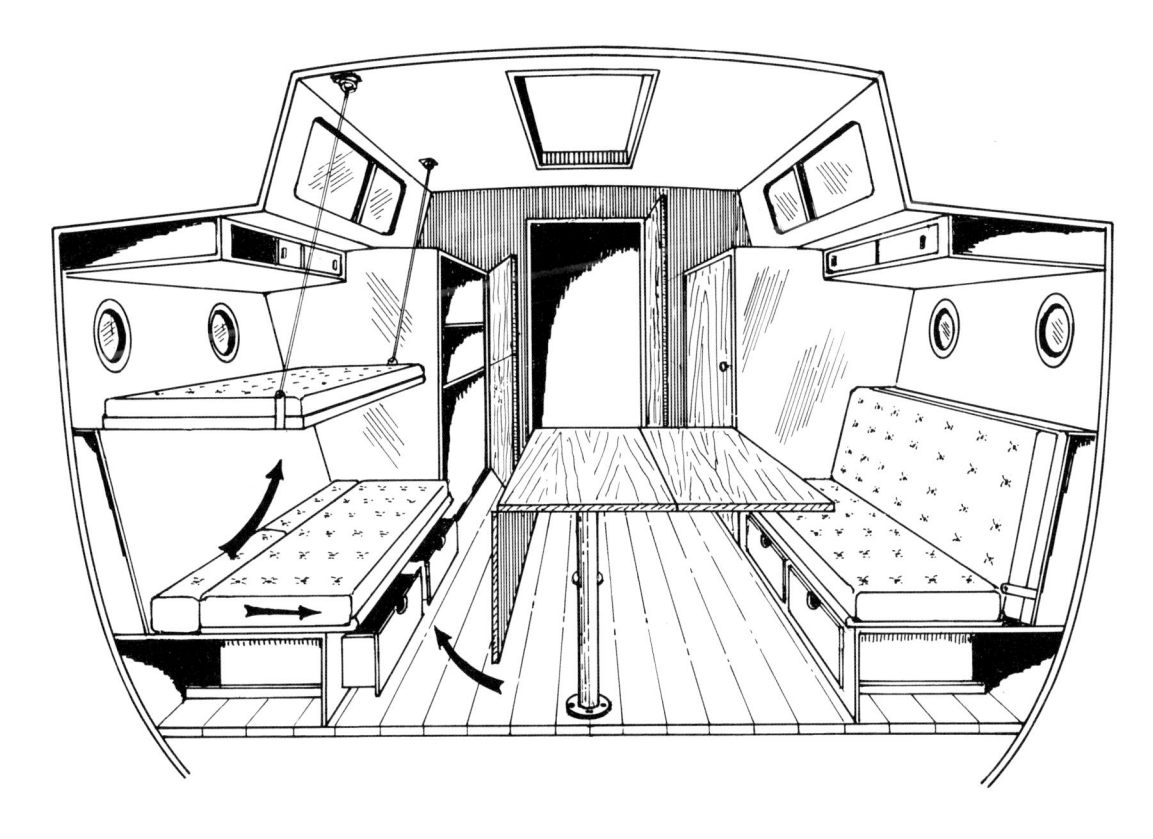

Fig. 17.5 Many space-saving devices can be incorporated in our self-design, as illustrated in the cabin view above.

select an appropriate lavatory basin, but otherwise there should be no great problems.

Large motor yachts may have a bridge or wheelhouse and attention needs to be paid to obtaining a good all-round view. A pleasing appearance within the general profile of the yacht can be obtained by correctly shaping the size of windows at bridges and cabins, and sloping the sides to provide a more rakish effect. Equipment such as wheel and column, radar sets and other navigational aids need to be located in their best position, along with a chart table and similar facilities.

Ventilation requirements should cater for the structure as well as for living spaces. A good flow of air through the bilges and to the frames and skin of the hull is needed. Accommodation areas will have vents, with (possibly) ducting to provide a flow and exhaust. If so, this could be tentatively schemed in the Plan and Elevation. The same applies to other compartments, while the galley and toilet may require forced ventilation to clear away nasty smells, which means a fan-type arrangement. The engine compartment certainly needs good ventilation to clear away fumes while providing a fresh supply of air. Vents vary in type, such as mushroom, cowl,

Plate 20 Navigational aids such as Sperry Marine's 340 CAS marine radar need to be located in the wheelhouse arrangement of our GA. (*Courtesy Sperry Marine.*)

and so on. You will have to select an appropriate type for your compartment and design.

Small sailing craft are usually steered by a tiller, but larger vessels and most motor yachts which have a wheel need a more elaborate steering arrangement between the wheel column and rudder. Two most common systems for small craft – bought off-the-shelf – are cable and rod-and-link. You may indicate the run of steering cable or rod in the Plan View and possibly a section in way of the wheel. The water and fuel supply system, electrics, battery, deck equipment and all other aspects concerned with the running and operation of the vessel need to be given thought – if only in speculative fashion – and indicated on the GA if possible. There is no great necessity for detail as this will come later, but some idea of space requirements and location is useful, as it means that major snags in detailed design are anticipated before they become a problem.

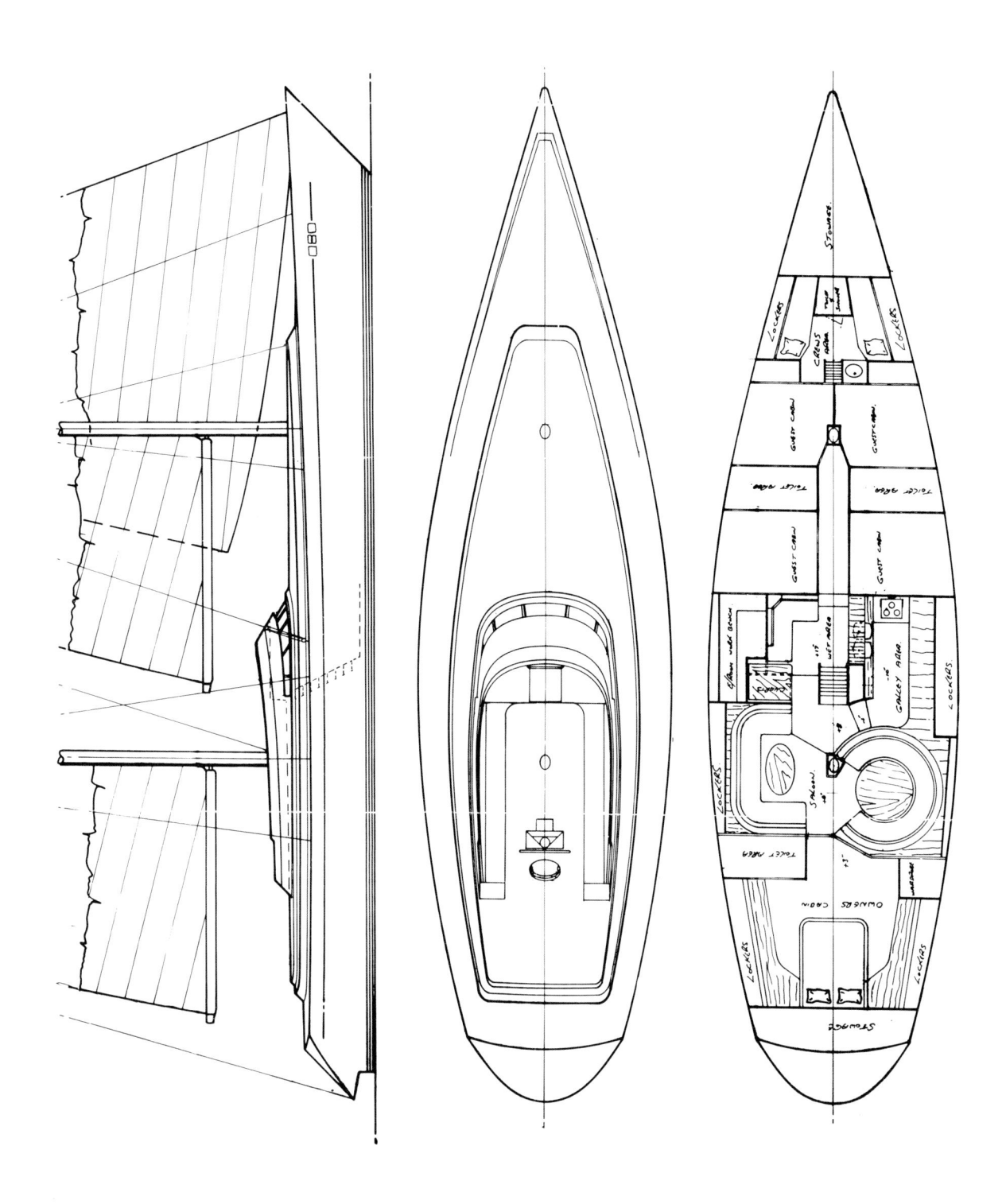

Plate 21 A typical General Arrangement plan for a yacht – the Ocean 81 by Southern Ocean Shipyard – is shown above. (*Courtesy Southern Ocean Shipyard.*)

It is impossible to describe all the various factors as they crop up; only the actual drawing of the plans will do this. The end result will be a design creation born of manufacturers' literature, consideration of structural aspects, employing ingenuity, checking space requirements and locating the gear and equipment to be fitted in the most appropriate places. The next task is to exactly delineate the geometry of the hull – the Lines Plan.

The Lines Plan is not just a geometrical description of the hull; it constitutes the efficiency and beauty of the vessel. To the naval architect it is both body and soul of design – as will be seen later. But first we must discuss the design aspects of the rudder.

18

A DESIGN HAND AT THE TILLER
Design of Rudder and Steering

Since time immemorial ships and boats have been steered by surfaces under the stern which are angled to produce turning moments or pressures. The stern rudder has remained one of the principal means of steering a vessel in the long history of sailing ships and powered craft, changing surprisingly little during that period. One departure was the steering oar, used by Nordic and Viking seafarers and even as far back as the Bronze and early Iron Age.

I was once on board the bridge of a very large cargo vessel during manoeuvring trials and when the captain ordered it to turn to starboard, the massive bulk of floating steel responded – slowly at first, then increasingly went into a turning circle, leaving a foaming circle in its wake. The wheel was then spun round for a reverse turn to port and it again edged round to trace a white figure-of-eight on the green sea surface. I remember the fascination I felt that such a large surface of steel could be controlled by a rudder surface so much smaller in size and area.

The complex mysteries of such turning feats lie in the kinematic forces involved, which resolve in the *turning force* when a rudder is set at a certain angle, usually no greater than 35° to be effective. Measure any rudder which is hard over, if you can, and you will find that this is the maximum angle that can be achieved. The fact that a rudder does act so effectively is suggested in Chapter 15, which discusses sail forces and design. In that chapter a rigid foil inclined at an angle to the flow was somewhat analysed, the situation being related to lift and drag forces that also occur in aerofoil theory and the lift forces on the wings of planes. It also mentioned that there is a pressure difference on either side of the foil which can be resolved in various ways. For a rudder, these force components are illustrated in Fig. 18.1.

Two things happen when a rudder is placed at a certain angle to the flow: there is a sideways drift *opposite* to it, then as the hull moves obliquely at this drift angle the hydrodynamic forces produced are very much more than that which the small area of rudder is capable of, and thus the hull slowly swings into a turning circle. Leonardo da Vinci commented on this rather amazing phenomenon, saying: 'Inasmuch as all the beginnings are often the cause of great results, so we may see a small, almost imperceptible movement of the rudder to have power to turn a ship of marvellous size and with very heavy cargo . . .'.

The creative hand that holds the tiller of rudder design is thus an

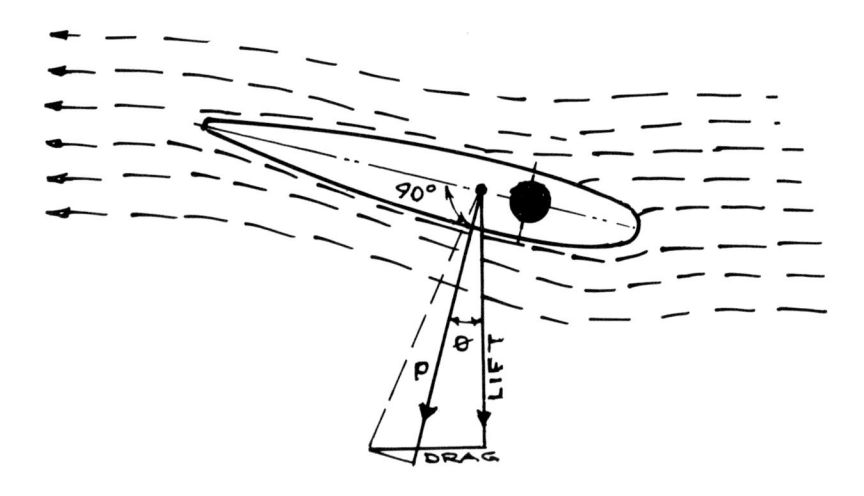

Fig. 18.1. A rudder has similar resolution of forces acting on it to that of a sail, except that water flow now produces the force components, as illustrated above.

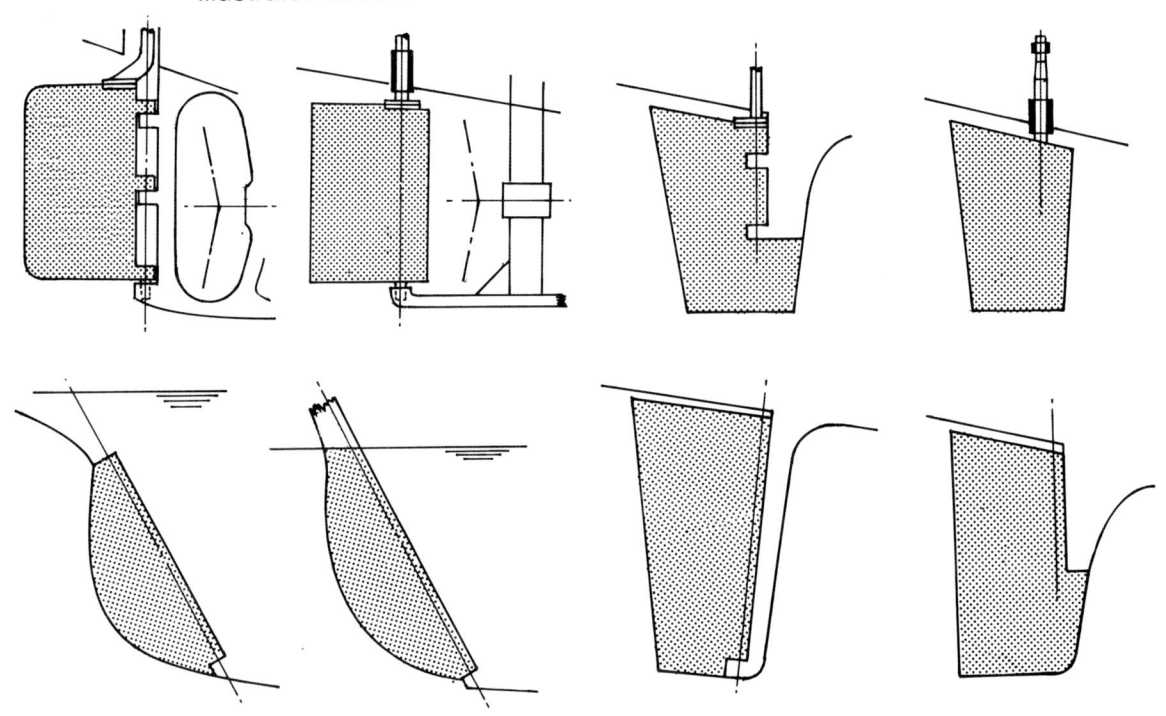

Fig. 18.2 There are various types of rudder and blade shapes, as shown in the different illustrations above. Such rudders can be balanced or unbalanced.

A Design Hand at the Tiller

important element in the total concept of yacht design. But first we need to appreciate some of the various types of rudders and their shapes. These are illustrated in Fig. 18.2. Such rudders can be balanced or unbalanced. The former is used to reduce the power needed to control it, but no more than 25 per cent of the total rudder area should be before the axis at which it turns or there may well be difficulties in bringing it back to the amidships position.

Without entering into the deeper waters of rudder theory, the relevant design points involved are:

(a) the position of the rudder should lie behind a propeller, if any;
(b) the largest possible area should be aimed for, compatible with other factors such as available aperture;
(c) the rudder should be deep and narrow rather than broad and shallow.

The reason why the rudder should be behind the screw is because it will benefit from the screw's race, thus enhancing its turning capability. Regarding the second point, a large area goes hand-in-hand with a large turning force while point (c), though not so obvious, depends on the fact that the deeper it enters the water, the more pressure is acting on it; not only that – for a narrow rudder this greater pressure acts closer to the axis of turn.

It now remains to find out how *much* rudder area is required. A general rule-of-thumb approximation is the simple ratio presented below and illustrated in Fig. 18.3

$$\text{Rudder area ratio} = \frac{A_m}{A}$$

where A_m is the underwater middle plane area and A the rudder area.

Acceptable values of the ratio vary from boat to boat. For small craft such as ours, the ratio for unbalanced rudders is about 1:15 to give a turning circle of between six and seven lengths of the yacht. For balanced rudders, something like 5 to 10 per cent should be added. Nevertheless, because so many factors such as shape, draught, etc. enter into the considerations, it

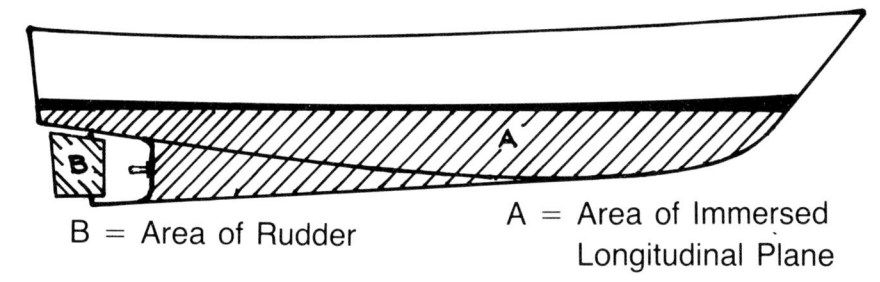

B = Area of Rudder
A = Area of Immersed Longitudinal Plane

Fig. 18.3 A rule-of-thumb approximation to find rudder area is to use a ratio incorporating this area with that of the underwater middle plane.

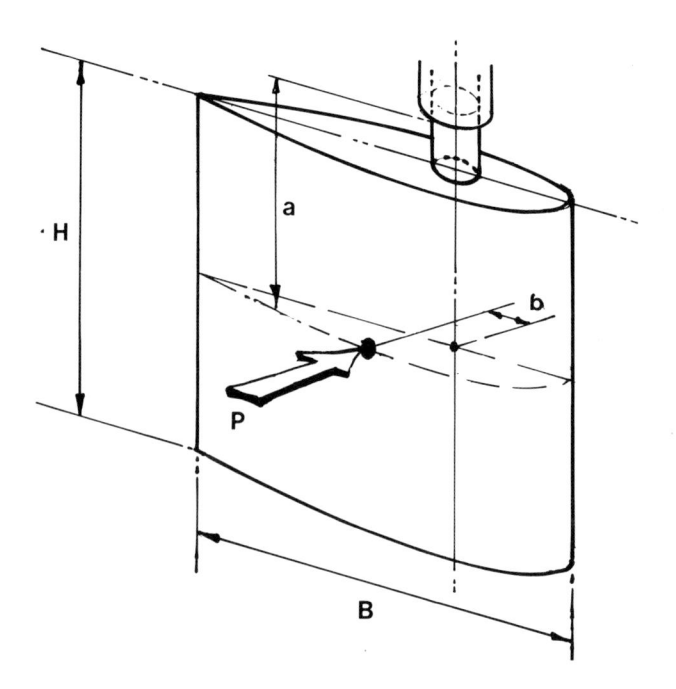

Fig. 18.4 To find rudder strength and diameter of rudder stock we calculate bending and turning moments. An idea of the forces acting is illustrated in the figure.

might be wise to obtain a ratio from the basis design and use this for your rudder area.

Regarding shape, it can be of a rectangular form or else curved. Rectangular type rudders are obviously easier to calculate, design and construct, but may not be suitable to the overall design. Curved shapes will require the use of Simpson's Rule or similar to calculate area. In either case, the edges or ends should not usually be left square but should be rounded or faired to reduce eddy drag, though square edges do improve rudder efficiency at large angles. As has been emphasised before, water abhors sharp edges and forms eddies around them.

There are three major points to take into account when considering the strength of rudder: its thickness, the turning moment or torque acting on it, and the diameter of rudder post or stock. Considering thickness, a good estimate may be made using the basis design, or else – as a rough guide – for flat rudders it can be 1.3 to 4 cm (about ½" to 1½") for wood and 0.3 to 1.0 cm (about ⅛" to ⅜") for metal.

The turning moment or torque on a rudder is found by multiplying the distance from the centre of pressure of the rudder to the axis to turn by the actual pressure (see Fig. 18.4). The pressure force is not easy to calculate, pressure varying with angle, speed of flow and shape of rudder. Two such reliable approximate formulae to obtain this force, with the rudder at 35°, are given by:

$$P = \frac{AV^2}{750} \quad \text{(with rudder behind propeller)}$$

and

$$P = \frac{AV^2}{900} \quad \text{(rudder not behind propeller)}$$

where pressure P is in tons, A is the rudder area in square feet, and V is the speed of the boat in knots. The position of the cp from the axis (b in Fig. 18.4) is generally assumed to be about ⅜ the rudder breadth, B, from the leading edge (the forward edge of the rudder) at 35°. Using this value, converted into inches, a formula to obtain torque (T) for balanced rudders is given by:

$$T = \frac{AV^2B}{500} \text{tons/inches}$$

where B is the rudder breadth in inches to the cp and the other terms are as before. This assumes that not more than ¼ of the whole area of balanced rudder is before the axis so that cp is not too far forward. For unbalanced rudders a safe assumption is to use the above formula for T, but multiply by 3. It may be noted that the vertical position of cp is *not* at the centre of area because pressure increases with depth of water. This need not concern us in our self design as the following approximate formulae will avoid any theoretical calculations involving the vertical position of cp.

The stock diameter at the head of the rudder can be constant all the way down, though it's quite usual to taper it slightly. This diameter is obtained by considering not only the turning moment, but also the bending moment $a \times P$ acting on the rudder (see Fig. 18.4). Now the distance a for the centre of pressure is not a straightforward calculation and is not located at the centre of area. For rectangular shapes the depth of cp is ⅔ the rudder depth, while for more complicated shapes such as curved outlines, more sophisticated calculation work is required. The diameter is obtained by considering the combined effect of turning moment and bending moment. For our DIY purposes we can use a formula – simple and crude though it may be – that should give reliable results. It is for rectangular rudders. It involves a constant K, based on the ratio of depth to breadth and applicable to most types of rudder supports used in boats and yachts. It can be applied to other rudder shapes, as explained later, and is given by:

$$d = \frac{KB \sqrt[3]{V^2}}{f}$$

where d is the diameter of the rudder stock in inches, B is the width of rudder in feet, V the speed of the yacht in knots and f the working stress of the material used in association with live loads and expressed in tons per square inch. Values of f can be found in any standard engineering or materials handbook, but here are some standard values for common boat materials: 5.5 for mild steel, 1.5 for cast iron, 0.35 for oak or Canada elm, 0.25 for teak or pitch pine, and 0.2 for fir or mahogany. Aluminium alloy or GRP would work out at about $1/3$ to $1/2$ the mild steel value. The K values for a representative sample of rudder types are presented in Table 18.1.

For curved shape rudders, take the ratio of the true area to that of a rectangular rudder with the same dimensions as the maximum depth and width of the rudder. The formula answer multiplied by this ratio should give a fair approximation.

To find the size of pintles, another simple formula is given by:

$$D_P = \frac{\text{Pressure on pintles} \times L}{2}$$

Table 18.1

Type of Rudder	'K' values		
Unbalanced Rudder Supported on 3 or More Pintles	0·40	0·42	0·45
Balanced Rudder – Supported Midway	0·44	0·49	0·54
Balanced Rudder – Supported at Bottom	0·49	0·55	0·62
Balanced Rudder – Supported inboard only	0·76	0·88	1·00
	2	2.50	3
	Depth to Breadth Ratio		

Interpolate for different Depth to Breadth Ratios.

Formula for Use: $d = \dfrac{K \times B \times \sqrt[3]{V^2}}{f}$

where d = Rudder Stock Diameter (inches)
 B = Rudder Breadth (feet)
 V = Boat Speed (knots)
 f = Working Stress, with Live Loads (tons/ins.^{2})

where D_p is the pintle diameter (in inches), and L is the length of pintles, also in inches. The pressure on the pintles (which is also in tons) can be taken as about ½P (total pressure) for bottom-supported rudders and those supported midway; about ⅜P for the bottom pintle of a rudder with three or more pintles (the others will have a lesser proportion of P); and P itself for those supported wholly inboard.

Finally, it may be said that Classification Societies such as Lloyd's Register of Shipping include in their rules all the relevant formulae and guidelines to obtain required sizes and dimensions of rudders, stocks and pintles, among other structural information.

19

SHIPSHAPE AND BRISTOL FASHION
Drawing the Lines Plan

There is an old shipbuilding expression which describes a well-built vessel as being '*shipshape and Bristol fashion*'. Bristol is a seaport city lying on the Bristol Channel and was once renowned for its excellent craftsmanship in marine construction and design. The steamship *Great Britain* – the first large iron ship, built under the direction of Isambard Kingdom Brunel – was launched at Bristol in July 1843 and has recently been resurrected from the Falkland Islands to return in triumph back to the city.

The essence of marine design lies in the hull shape; it needs to be what is termed hydrodynamically smooth. Such a hull offers the least resistance to motion in water, but the difficulty in drawing such a smooth shape – assuming the form characteristics have been determined – is the fact that most hulls have curvature in *two* directions: longitudinally, or along their lengths, and transversely, from deck to keel.

Once, when I was in charge of a small yacht design office on the South Coast – a very cramped place cluttered with the usual long boards and strewn drawings – I entrusted the task of drawing a Lines Plan to a junior designer recently out of his traineeship. He meticulously made his preparations, leaving time for the cartridge paper to stretch, then went ahead with the drawing. A few days later I wandered by to see how he was progressing.

'You have,' said I, 'drawn some very nice curves.'

His chest puffed in peacock pride.

'Thank you, Sir.'

'You're sure they're accurate?'

'Oh *yes*, Sir. I've checked them very carefully.'

'Well, to be *absolutely* sure, I'll double-check.'

A few moments later I gave him the famous stare cultivated by all chief designers.

'There seems to be a discrepancy of about half-an-inch in places,' I growled.

His expression reminded me of a stuck pig – and the days when *I* had made the same mistake and scared response. But behind the tough timbers of those in the marine design business there lies a soft cargo. My Boss had gently asked whether I had kept my pencil well-sharpened when marking off the dimensions and transferring them to the other views. I asked him the same question.

Swallowing tightly, he produced a stub of pencil as blunt as the bows of a Thames barge, which led me to lecture him on how the thickness of a pencil mark can drastically affect the accuracy of a Lines Plan. When drawing the plan you should constantly bear in mind any factors that may detract from such accuracy, as they will not only affect the performance of yacht, but also every other drawing.

Early shipbuilders used eye and judgement to obtain their hull shape; then ship plans began to appear (the earliest date from the 14th century), along with written instructions. Master shipwrights began to create desired hull shapes in mould lofts, even – like Phineas Pett – curving the lines beyond the waterline and producing hydrodynamic forms in the sense of modern naval architecture. Progress reached the stage where the three-dimensional curvature of a hull could be shown on a two-dimensional sheet of parchment or paper – to be first called the Draught Plan, then later, the Lines Plan.

The first step in drawing this plan is to make a grid comprising the waterlines, bow and buttock lines, and stations. On transparent paper or linen the grid is best drawn on the reverse side, as erasing can be done without affecting the lines, but obviously this is not possible on thick paper of the cartridge type. In this case it might be best to draw it in ink, but remember to keep the lines very thin.

The scale of the drawing needs to be as large as possible, say from $1/10$ up to $1/5$ for metres ($1/5$ may be too large for large yachts), and 1″ up to 3″ to the foot for Imperial units. The size of drawing sheet should be in the region of a metre or more (three feet and above) to incorporate the three views of Elevation, half Plan View and Body Plan.

Plate 22 Early shipbuilders used eye and judgement to obtain their hull shape. Master shipwrights began to create desired hull shapes in mould lofts, such as above, which is an early photo of Thornycroft's mould loft at Southampton, Hants.

To start the plan we draw a baseline midway on the paper, extending it right across the sheet, a centreline (℄) near the bottom and a vertical centreline at the top right hand corner, placing this to allow for both sides of yacht to be drawn in section. It is from these three datum lines that the grid – and, indeed, the rest of the plan – will develop.

On the baseline we then measure off ten equal intervals for the stations. The 11 stations thus produced should extend all the way from the deck in profile down to the centreline in Plan, and should be labelled 0 to 10, beginning with the AP. Eleven stations are integral to the displacement and other calculations (21 in large vessels) and it might be worth adjusting the floating waterline length slightly in order to obtain a spacing that is simple and easy to handle. For instance, a 9.8-metre waterline could be extended to 10 metres to obtain 1-metre spacing. Half-stations (even quarter-stations) are usually drawn at the ends to assist the fairing process and allow for greater accuracy in the calculations.

Above the baseline we then measure off and draw the waterlines (W/Ls), according to the draught of the vessel. The spacing needs to be reasonably close, say between 15 and 22 cm (about 6″ to 9″), depending on the size of craft. Above the floating waterline are then drawn the Level Lines, which can be spaced wider and go up to the deck. The wider spacing for Level Lines is because most hulls are relatively 'straight' in the curved sense, above the water, while at the turn of bilge and below, we should subdivide the W/Ls with half waterlines in order to aid the fairing process in this region. The W/Ls and Level Lines are then numbered W/L 1, W/L 2, etc., beginning with the first above the baseline, and should be extended right across to the Body Plan view.

Next, we measure off and draw the Bow and Buttock lines in the half Plan View, adopting a similar spacing to that for the W/Ls, and once again choosing simple divisions based on the half beam. The nearest one or two lines away from the centreline are subdivided to enable the fairing process in this region, and the lines are numbered from the centreline. The Bow and Buttock lines are then drawn as vertical lines on the Body Plan view, being extended to the deck.

We now have a grid of lines in all three views and in the Elevation part we draw the profile as per our initial GA plan. In the semi-Plan we outline the half-deck as taken from that plan. At this point we can refer to Fig. 19.1, which also shows how one of the curves is drawn and then transferred to the other views, as part of the fairing process that develops the Lines Plan.

The moment of truth has now arrived; we have a framework and bare outlines of our design. It now needs to be 'fleshed' in with a series of curves that are smooth and accurate in all views and correspond to the form characteristics previously obtained.

It will be remembered that a very crude and general outline of the hull shape was obtained, based on the form characteristics, as illustrated in Fig. 14.3 in Chapter 14. It is here that we start the Lines Plan process by transferring the sketched half-section onto the Body Plan part of the grid.

Shipshape and Bristol Fashion

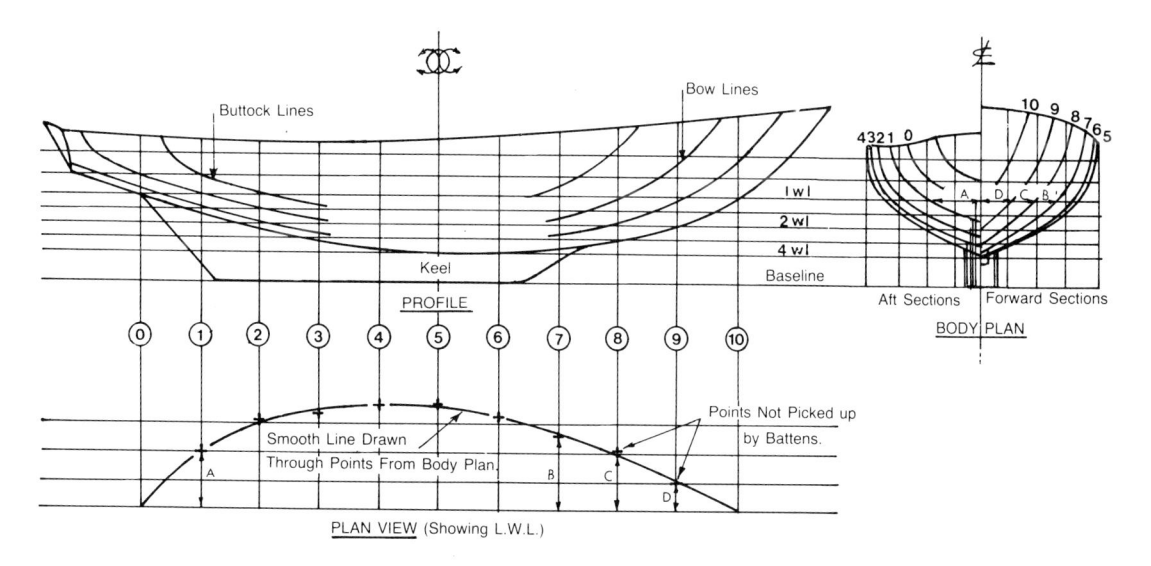

Fig. 19.1 A grid is first drawn to develop the curves of a yacht Lines Plan. These curves are transferred to other views as part of the fairing process, until they correspond in all views. The above figure shows how this is done.

The sheerline heights and those of the keel at each section are then marked on this view so we have endings for the intended half-sections to be sketched, together with half-breadth dimensions at deck (taken from Fig. 14.3) at each station. Where the sheer height and half-breadth at deck intersect, this will be the terminal point of the half-section at the deck-at-side. Naturally, the keel heights will be marked on the centreline of the hull. Knowing the boundaries of the half-section, we can fill in with one more dimension – the half-breadth at LWL, taken from Fig. 14.3.

It is at this stage that we do more freehand sketching, using the sparse points so plotted to draw the likely shape of sections at each station, making sure there is a certain graduated continuity in the set of curves sketched so that longitudinal smoothness is somewhat approximately maintained. Remember, forward half-sections are drawn on the right of the centreline and after ones on the left. Waterlines now need to be drawn using these rough half-sections and this is done by use of a strip of paper upon which are marked the half-breadths at each level and waterline, which are then transferred to the Plan View (see Fig. 19.2a). The result is a set of points at each station for each level and waterline. A smooth line is drawn through the relevant points pertaining to each level and waterline, as illustrated in Fig. 19.1. These curves are drawn using either battens and weights or else sweeps and other plastic, curved instruments. The vertical dimensions of the half-sections are then lifted off at each bow and buttock line to produce smooth curves in the Elevation in similar fashion to those in the Plan View.

137

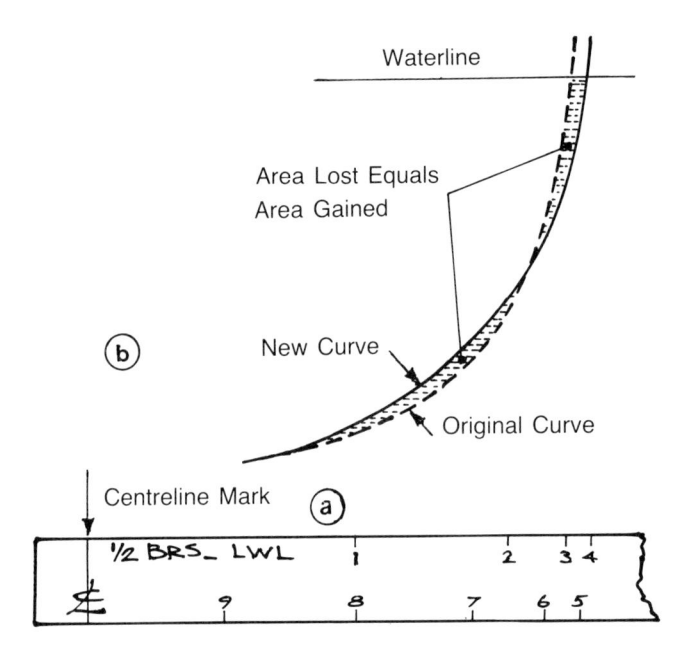

Fig. 19.2 Strips of paper are used to transfer points of a yacht curve to other views. At the same time we must consider our design displacement and balance the various curves of our Lines Plan so that any area lost equals area gained, in order that the design displacement is maintained.

Obviously, because all the points will not have been picked up – and it would be a million-to-one chance if they had been – there has to be a readjustment to the half-sections in that view, transferring the new dimensions so they will correspond with the other two views.

The new points of half-section have to be faired into smooth curves and transferred back to the other two views, which may again have to be modified. This process of adjustment and readjustment between the three views is continued until they are in *exact* correspondence with each other at every station, waterline and level line, bow and buttock line, and at the same time produce smooth and fair curves in each plane. Drawing the Lines Plan sounds a tedious process – and it is. But once you are into the exercise it actually turns out to be quite enjoyable – especially when you have achieved a curve that not only corresponds in all views, but complies with the naval architect's rigorous demands of fairness and smoothness.

The Lines Plan story is not yet complete; we have to deck the sections. Most decks will have a camber of some form, usually rounded. A common figure for yacht camber is about 1:30 to 1:25 (about ⅜" to ½" to the foot). Camber is normally part of the arc of a large circle, though parabolas can be substituted with hardly any difference. For our DIY purposes we can find the height of camber at a certain half breadth by simply finding the height of camber at the deck-at-centre then using a sweep or other template, draw a

Shipshape and Bristol Fashion

Plate 23 Long curves in the Plan and Elevation of the Lines Plan can be drawn using weights and battens then sighting along their length to obtain fairness. Pear shapes are ideal for the Body Plan.

curve to the deck at side. For example, if the half breadth were 1.5 m (or 150 cm), using 1:25 we would have a height of camber above deck at side of 6 cm. Starting at amidships and using the same sweep, we can find the camber heights at each station to be drawn on the Body Plan. These heights of camber – taken from the baseline – can be plotted at each station in the elevation and a smooth curve drawn to give the deck at centreline. Little or no adjustment should be necessary – so we now have a complete outline of the hull. More accurate methods of drawing camber are illustrated in Fig. 19.3., which shows a graphical (a) and mathematical (b) approach.

Plate 24 Readjustment between the three views on a Lines Plan should eventually arrive at correspondence and a smooth, fair hull such as Cougar Marine's 12-metre yacht seen in plan, above. (*Courtesy Cougar Marine.*)

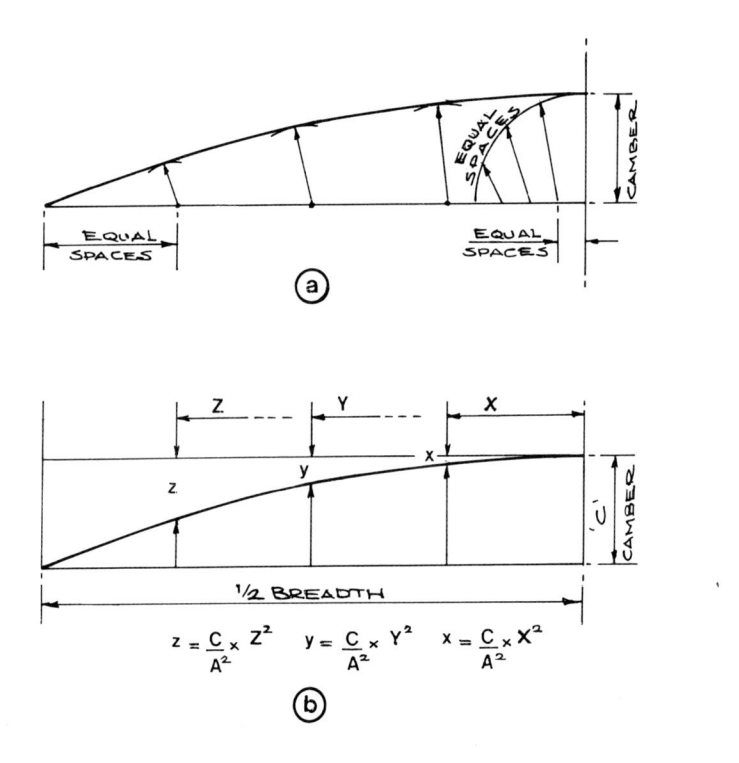

Fig. 19.3 Accurate ways to determine our deck camber are illustrated by the (a) graphical and (b) calculated approaches illustrated above.

$$z = \frac{C}{A^2} \times Z^2 \quad y = \frac{C}{A^2} \times Y^2 \quad x = \frac{C}{A^2} \times X^2$$

It may be that the Lines Plan will still have to be slightly adjusted when the detailed calculation for displacement is made, and this will require a modification of shapes so that any area lost = area gained, to balance the required displacement, see Fig. 19.2(b), but the major part of the task is now completed and all else is a tinkering process. We now need to tabulate the various dimensions in what is called a *Table of Offsets* in order that the relevant calculations may be performed. It may be emphasised that these dimensions or offsets are to the *inside* of skin for metal boats, and to the *outside* of skin for GRP and wooden craft.

So the hull form now has flesh and framework; there is still a long stretch of the spiral voyage to traverse before we reach our terminal port of design. Factors that will ensure a successful cruise are that our creative yacht is strong, has a heart of oak and timbers that do not shiver in turbulent seas – the next topics in this design voyage.

As an aid to the fairing process diagonal lines, as mentioned in Ch 2, are drawn in the Body Plan. Such lines are drawn from the intersection of centreline and LWL and through the sharpest curvature of bilge. The points picked out at each station are plotted on the Plan View and must also obey the rigorous demands of fairness and correspondence.

20

BUOYANT AND AWEIGH

Calculation for Displacement, Weight, and Centres of the Design

We now come to more calculations, but please believe they will not be a mathematical dose of castor oil.

Legendary to scientific history is the well-known tale of how Archimedes jumped out of his bath and ran naked through the streets of Syracuse shouting *'Eureka!'* (I have found it!).

Archimedes was born in Syracuse in 287 BC and slain 75 years later during a siege of the town. During the intervening years he proved to be a versatile genius, being the first to apply scientific thinking to everyday problems, giving proofs for finding areas, volumes and the centres of gravity of circles, spheres, curves and spirals, as well as inventing engines of defence for Syracuse against the Romans. He is credited with the invention of the *Archimedes Screw*, a cylindrical device for raising water.

How does an ancient scientist enter into the world of yacht design? Well, his 'Lady Godiva' act was prompted by a brilliant idea concerning floating bodies that supposedly came to him when soaking in his tub. Simply stated, his idea was as follows:

The weight of a body floating in water equals the weight of water it displaces.

At the time, Archimedes was more interested in finding a way to accurately weigh gold in order to avoid any discrepancies – honest or otherwise – than in establishing this most important principle of naval architecture. For our purpose, its relevance lies in the fact that the *underwater volume of a hull*, multiplied by the density of water, will give our yacht displacement.

My first encounter with displacement calculations was when a designer in the office where I trained wanted me to 'run round' some half-sections of a Body Plan with a planimeter (it could have been an integrator – I don't remember). These instruments are very useful in finding such areas, and eradicate the somewhat tedious manual calculations involving Simpson's Rule, as explained in Chapter 12. The answers are read off a cursor and have then to be multiplied by a constant provided in the instructions, dependent on the scale used. All that is required is a steady hand and nerve. And in my youthful and inexperienced state, that was the catch!

My adolescent hands managed to set up the lever bar and mainframe in the right position for circuiting the curves, then fingers began nervously guiding the pointer round one of the sections. The longer I traversed its

Design Your Own Yacht

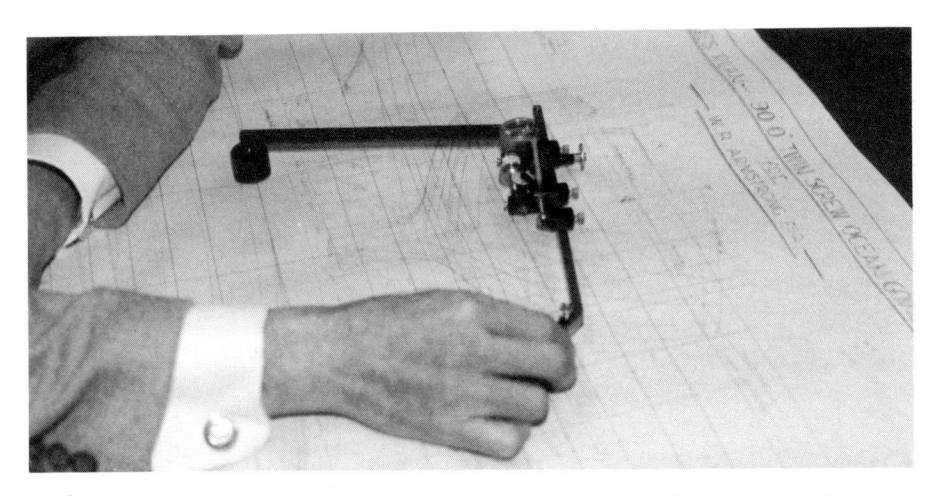

Plate 25 A planimeter may be used to perform displacement calculations for our self-design.

contour the more 'wobbly' grew my efforts. Near the end I suddenly froze – literally – and took a few deep breaths. Thank God! No-one else in the office paid attention to my icicle posture – though one old hand turned a quizzical eye in my direction. I reassured myself that no bombs would fall

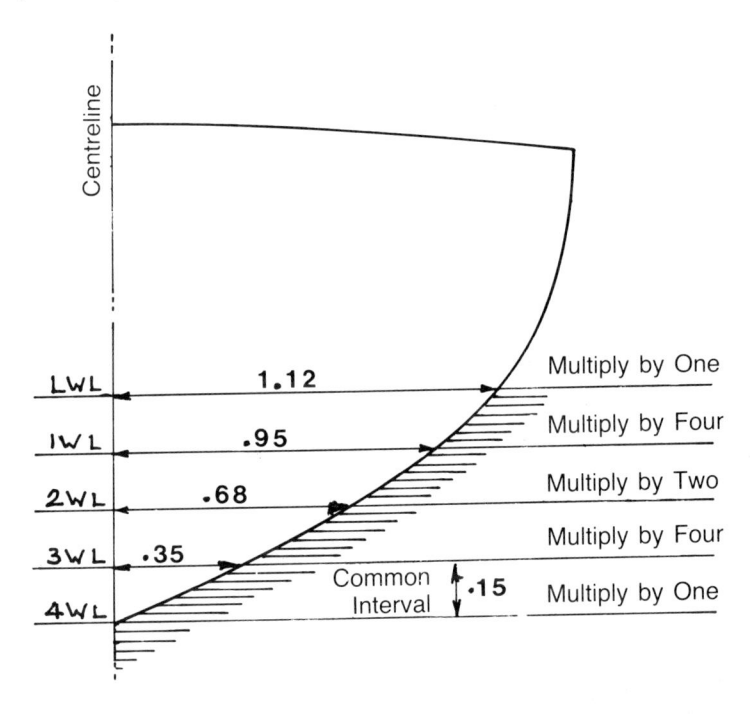

Fig. 20.1 The above figure illustrates the calculation of underwater sections of our yacht design in order to obtain displacement.

Buoyant and Aweigh

through the roof if my first effort was inaccurate, and completed the graphical journey, feeling at the end as if I had run a mental marathon.

Surprisingly enough, any inaccuracies due to the 'wobbling hand' effect were minimal, and were ironed out by taking a second circumnavigation of the section, then taking an average of the two readings. A planimeter or integrator is rather expensive for our DIY purposes, so assuming we have drawn our Lines Plan, we can apply Simpson's First Rule to each half-section in the following manner to obtain our displacement.

Figure 20.1 shows a typical half-section through a hypothetical yacht design and – using our knowledge of curved areas from Chapter 12 – we can put the ordinates or offsets shown through Simpson's Multipliers as presented in Table 20.1. The end result will be the half-area of section and such half-areas at each station are again put through Simpson's Rule in tabular form, as shown in Table 20.2, to eventually arrive at the volume of displacement. This volume now only needs to be multiplied by the density of water to obtain the actual displacement of the yacht. Table 20.2 gives the density of water (fresh and salt) for both systems of units.

As an additional aid and check, if the half-areas are plotted on a base of waterline length (see Fig. 20.2), they should produce a curve of areas as previously discussed that should be smooth and fair. This ensures your Lines and Body Plans are also reliably smooth and fair. One more point: *do* apply the correct water density to your volume of displacement (river boats should use fresh water densities) – and remember to multiply by two for both sides of the hull. I confess to having forgotten it myself, and it can be quite disconcerting to find you have an apparently featherweight displacement.

Table 20.1

Col 1	2	3	4	5	6	
LWL	1·12	1	1·12	0	0	Col 1 is Ordinate No. (W.L.)
2WL	0·95	4	3·80	1	3·80	Col 2 is Ordinate ½ Breadth
3WL	0·68	2	1·36	2	2·72	Col 3 is Simpson Multiplier
4WL	0·35	4	1·40	3	4·20	Col 4 is Function of Areas
5WL	0	1	0	4	0	Col 5 is No. of CI's from LWL
Totals			7·68		10·72	Col 6 is Function of Moments (Col 4 × Col 5)

CI = 0·15

$$\tfrac{1}{2} \text{ Area} = \frac{7\cdot68 \times 0\cdot15}{3} = \mathbf{0\cdot384 m^2}$$

$$\text{V.C.G.} = \frac{10\cdot72 \times 0\cdot15}{7\cdot68} = \mathbf{0\cdot21\ m} \text{ Below LWL.}$$

Table 20.2

Col 1	2	3	4	5	6	7	8	
0	0	1	0	5	0	0	0	
1	·080	4	·320	4	1·28	·07	·022	Col 1 is Station No.
2	·210	2	·420	3	1·26	·10	·042	Col 2 is ½ Area of Station
3	·325	4	1·300	2	2·60	·15	·195	Col 3 is Simpson Multiplier
4	·390	2	·780	1	0·78	·19	·148	Col 4 is Function of Areas
5	·394	4	1·576	0	0·00	·21	·331	Col 5 is No. of CI's from Sta. 5
					5.92			
6	·355	2	·710	1	·71	·20	·142	Col 6 is Function of Moments
7	·280	4	1·120	2	2·24	·17	·190	Col 7 is C.G. of Sta. below LWL
8	·160	2	·320	3	·96	·12	·038	Col 8 is Function of Moments
9	·055	4	·220	4	·88	·10	·022	
10	0	1	0	5	0	0	0	
Totals			6·766		4·79		1·130	

CI = 0·65m

$$\text{Volume} = 6 \cdot 766 \times 0 \cdot 65 \times \frac{2}{3} = \mathbf{2 \cdot 93 \; m^3}.$$

$$\text{LCB} = (5 \cdot 92 - 4 \cdot 79) \times \frac{0 \cdot 65}{6 \cdot 766} = 0 \cdot 109 \; m \; \textbf{Abaft St. 5}$$

$$\text{VCB} = \frac{1.130}{6 \cdot 766} = \mathbf{0 \cdot 167 \; m} \; \textbf{Below LWL}$$

Displacement = Volume of Main Hull (as above) + Volume of Appendage (Keel)
= 2·93 m³ + 0·31 m³ = 3·24 m³ or 3·24 × 1·025 = **3·32** tonnes
(1·025 tonnes salt water = 1m³)

$$\text{LCB} = \frac{(\text{Vol. of Main Hull} \times \text{LCB}) + (\text{Vol. of Appendage} \times \text{LCB})}{\text{Vol. of Main Hull} + \text{Vol. of Appendage}}$$

$$= \frac{(2 \cdot 93 \times 0 \cdot 109) + (0 \cdot 31 \times 0 \cdot 80)}{2 \cdot 93 + 0 \cdot 31} = 0 \cdot 175 \; m \; \text{Abaft St. 5}$$

$$\text{VCB} = \frac{(\text{Vol. of Main Hull} \times \text{VCB}) + (\text{Vol. of Appendage} \times \text{VCB})}{\text{Vol. of Main Hull} + \text{Vol. of Appendage}}$$

$$= \frac{(2 \cdot 93 \times 0 \cdot 167) + (0 \cdot 31 \times 0 \cdot 218)}{2 \cdot 93 + 0 \cdot 31} = 0 \cdot 218 \; m \; \text{Below LWL}$$

Table 20.3

Fresh water	one ton	36 ft³
	one tonne	1·0 m³
	one foot³	62·4 lbs
	one metre³	1·0 tonne
Salt water	one ton	35 ft³
	one tonne	0·975 m³
	one foot³	64 lbs
	one metre³	1·025 tonne

Concerning the offsets used, it may be emphasised again that these are to the *inside* of skin; the overall displacement can always be evaluated by making allowance for the skin thickness. For GRP and wooden yachts the displacement calculation is always carried out using offsets to the *outside* of skin.

For those readers who are fortunate enough to own a yacht already – yet do not know its exact displacement – it can be calculated by lifting the offsets with the use of measuring devices and battens as illustrated in Fig. 20.3. The process requires a degree of patience and accuracy, plus the fact that you will have to draw the 'lifted' offsets on paper to ensure their fairness. The calculation procedure is exactly the same as described previously.

Having obtained the displacement of design from the Body Plan, it should comply fairly closely with your initial estimates. If there is a large – and

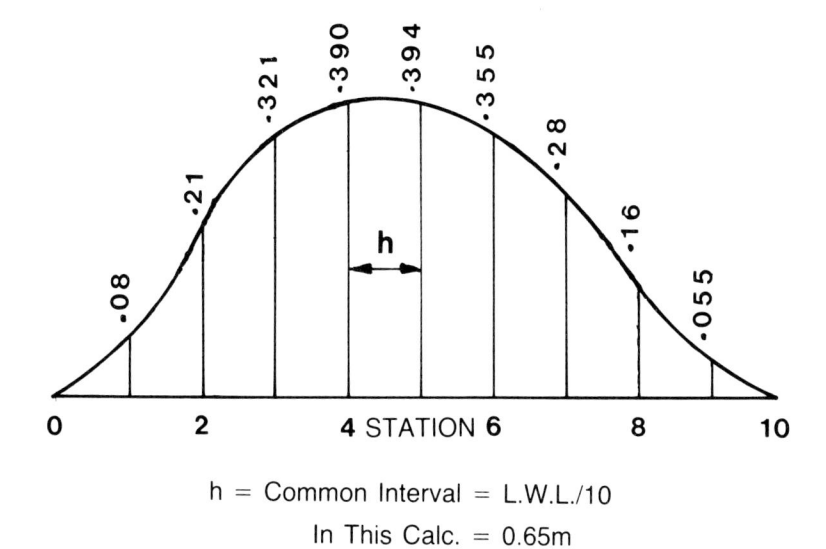

h = Common Interval = L.W.L./10

In This Calc. = 0.65m

Fig. 20.2 The areas of half-sections of our underwater hull can be plotted to obtain a Curve of Areas as shown above. The smoothness of this curve ensures the hull is reliably fair.

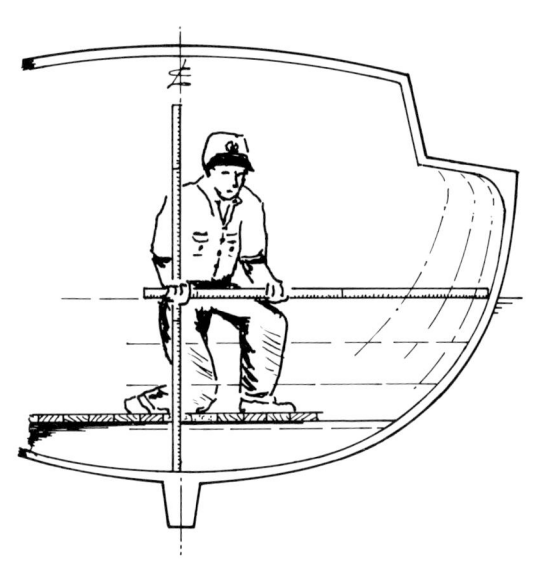

Fig. 20.3 For an existing yacht – if no other data are available – the displacement may be obtained by lifting offsets as shown above, using measuring devices such as battens and measuring tapes.

unwanted – discrepancy, then check your initial estimates and the actual calculation. If these are correct in themselves, then you can only accept that the initial estimate was not too accurate (in which case there is no adjustment to be made), or that you have to alter your preliminary Lines Plan as drawn, to suit the estimated displacement. This latter course is usually the one taken because the estimate was based on geometric coefficients.

There are two ways in which the Body Plan at this stage can be modified – as well as using a combination of both. The first is to alter draught (if it is not a fixed dimension); the second is to alter the shape of the body sections. If we change the draught then there is a certain formula that will give us a good estimate of how much parallel sinkage or rise is required to obtain the new displacement. It is called the *Tons per Inch* (TPI) in British units and the *Tonnes per Centimetre* (TPC) in metric units. Without showing how it is derived, the formula is given by:

$$\text{TPI (tons)} = \frac{A_w}{420} \qquad \text{For salt water } (A_w \text{ is in sq. ft.})$$

and

$$\text{TPC (tonnes)} = 0.01 A_w \quad \text{for salt water } (A_w \text{ is in m}^2)$$

We can convert from one to the other by the following formula:

$$\text{TPC} = 0.40 \text{TPI}$$

The above formula assumes a weight is added (or taken off) at the cb, and also that the resultant layer is not very thick, as should be the case for us. After all, if there *is* a wild discrepancy it would pay to go back over all the calculations and very carefully check to see if we haven't made some fundamental mistake or wrong assumption in our initial design. Once we have found the amount of rise (or fall) required to adjust for the correct displacement, we need to re-calculate to ensure we obtain a close approximation to the required displacement. As an example in using the formula, suppose A_w is 20 m^2, then TPC is $0.01 \times 20 = 0.2$ tonnes. If we needed to add 0.5 tonnes to our displacement then the waterline would have to rise $0.5 \div 0.2 = 2.5$ cm.

Should the calculated buoyancy not comply with that obtained at the preliminary design stage then, adopting the second approach, the sections need to be modified to obtain either more or less volume of displacement, depending on whether there has to be an addition or reduction. This is done by filling or reducing the section curves, using judgement in the process, as illustrated in Fig. 19.2.

The next part of the buoyancy calculation is to locate the longitudinal and vertical cb (remember, a formula was given in Chapter 14 to obtain an initial position for vcb). Using the calculation in Table 12.3, Chapter 12, as a basis, Column 7 showed that levers were used to obtain a horizontal position of area, and Column 5 used ½ ordinate^2 for the vertical position. Although we are now dealing with volume, the ordinates for station areas could be plotted to give a Curve of Areas, so the same principles apply for lcb. In Table 20.3, Column 5 corresponds to that in Table 12.3. Notice that the levers start from amidships and the separate moments obtained are subtracted to eventually obtain a position either aft or forward of the midship station, depending on which moment is larger. For vcb we cannot use the same principles because we are dealing with areas and their CG at each station. In this case we find the centroid of each area from the LWL and use these as levers. See Columns 4 and 7.

Once again, if the calculated position of cb is not quite where we desired (say, more aft), then – without tampering with the displacement – we would have to fill out the after sections and reduce the forward ones until the required shift has been obtained. This is very much a trial-and-error procedure that requires recalculation every time the curves have been changed. But you will soon get a 'feel' for how much to change and should quickly arrive at a reasonable answer within two or three trials. If the vcb needs to be altered then obviously one could shift the waterline up or down (if it were so critical), or else make sectional changes that will affect a movement in the right direction. It may be that the lcb is also affected, so should we require a double shift, it is best to incorporate the change of sections with this in mind. In truth, the vcb is not as critical as the vcg, which can be altered by shifting weights or adding ballast.

Weight and cg are the next part of our calculation.

In the DIY sense of our design – and considering yachts and motor cruisers are small compared to other vessels – the approximate formula and approach explained in Chapter 14 will probably be adequate for our purposes. As an alternative, I propose a somewhat simplified approach and make a straightforward comparison with the basis design as follows:

$$\text{Design total weight} = \frac{\text{Design displacement}}{\text{Basis design displacement}} \times \text{Basis total weight}$$

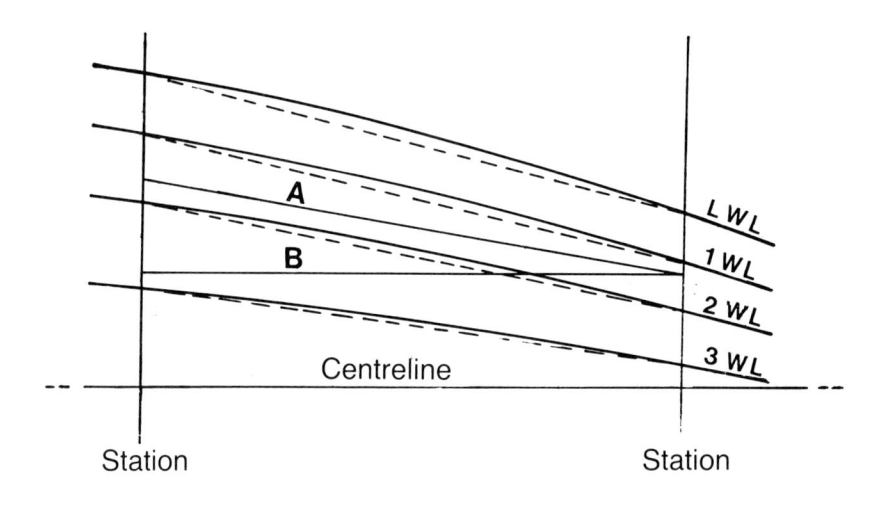

Fig. 20.4 Adjustment for longitudinal curvature of a developed hull is done by taking the ratio of mean slope between stations, as illustrated above.

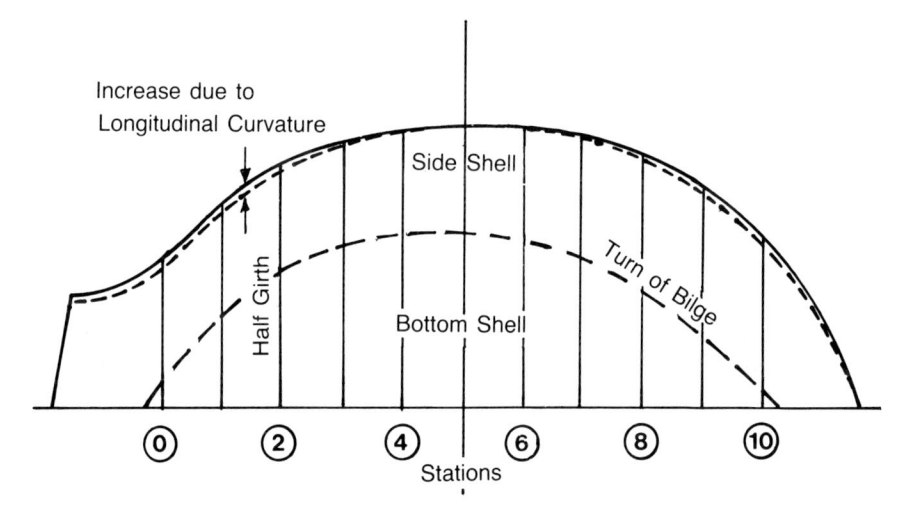

Fig. 20.5 The 'flattened' hull depicted above, is a means of obtaining hull weight and cg for our design.

But this does not solve the problem of locating the cg. The proper calculations are very lengthy, but unavoidable, so I will try to reduce the procedure to its bare essentials and make some gross simplifications in the process. The hull weight is a major part of the total, so we need to somehow make it amenable to calculation. This is done by running a strip of paper round the half-girth at each station to find the curved length. This only takes account of the *transverse* curvature, so to allow for *longitudinal* curvature we start at the midship station and find a mean slope of waterlines between the stations, as shown in Fig. 20.4. The modifying ratio A/B multiplies the half-girth in that region (the midship half-girth remains untouched) to obtain a 'flattened' hull as shown in Fig. 20.5. But the thickness of a hull is not constant; the sides are thinner than the bottom up to the turn of the bilge, while thickness tapers off towards the ends. To simplify matters we can divide the 'flattened' hull into two, separating the side from the bottom, as indicated in the figure by the dotted line, which is drawn for half-girths up to the turn of the bilge.

We now need an estimate of thickness for the upper and lower portions of the curve. This can be done by either basing them on the guidance design we chose, using the rules of a Classification Society such as Lloyd's Register of Shipping, or else making our own calculations, as briefly described in Chapter 22. To allow for tapering thicknesses towards the ends we can judiciously select a constant thickness for each half of curve, then, putting the part half-girths of each through Simpson's Rule, the area and thus weight for each portion can be obtained to give the total weight of hull skin (see Table 20.4). But this does not allow for its skeleton framework of frames, longitudinals, etc.

The lcg of hull skin is calculated as before, using levers, as shown in Table 20.4, but the vcg requires that we estimate the cg of the curved shape of each section, based on the graphical technique shown in Fig. 12.4, which is self-explanatory. Because hull sides are sufficiently close to being straight, for our purposes, we can 'spot' an estimated cg for each station and measure the distances from LWL. The bottom can follow the procedure of Fig. 12.4 and then again, the cg distances obtained can be measured from the LWL. Such distances are applied in the same way as for vcb, as shown in Table 20.2.

For transverse frames the mean half-girth of a group of similar frames can be measured, then multiplied by the number of frames. This group weight can be found either by direct calculation or using data from materials handbooks, which give the weight per foot or metre of such sections. The group cg can be determined using the graphical method mentioned previously. Other scantlings such as the overhanging ends of hull, deck, longitudinals, bulkheads, etc., can be calculated using similar techniques to those already described, while the superstructure and cabins could even be approximated to rectangles or other simple shapes. Simpson's Rule, of course, is constantly used in the whole operation, which would take too long to describe in detail, but I trust you would now have enough confidence to

Table 20.4

Sta.	½ Girth	SM	Area funct.	Lever	Momt funct.	VCG	Momt funct.
0	0·15	1	0·15	5	0·75	0·92	·14
1	0·70	4	2·80	4	11·20	0·76	2·13
2	1·00	2	2·00	3	6·00	0·70	1·40
3	1·20	4	4·80	2	9·60	0·65	3·12
4	1·25	2	2·50	1	2·50	0·62	1·55
5	1·30	4	5·20	0	0·00	0·60	3·12
					30·05		
6	1·25	2	2·50	1	2·50	0·62	1·55
7	1·10	4	4·40	2	8·80	0·65	2·86
8	0·80	2	1·60	3	4·80	0·68	1·09
9	0·50	4	2·00	4	8·00	0·75	1·50
10	0·10	1	0·10	5	0·50	0·92	·09
Totals			28·05		24·60		18·55

CI = 0·65

Area (both sides) = $28 \cdot 05 \times 0 \cdot 65 \times \dfrac{2}{3}$ = *12·16 m²*.

Weight (at 15 kg/m²) = **0.18** tonne.

LCG = $\dfrac{(30 \cdot 05 - 24 \cdot 60)}{28 \cdot 05} \times 0 \cdot 65$ = **0·13 m** Abaft Sta. 5

VCG = $\dfrac{18 \cdot 55}{28 \cdot 05}$ = **0·66 m** Above Base Line

apply such techniques, using initiative and ingenuity, where necessary. For instance, it is quite acceptable to make an intelligent guess in placing the cg of an awkward part of a hull structure and outfit. Finally, we should add about 5–7½% to the hull and structure weight to allow for such items as connections, fastenings, paint, and so on, which cannot easily be calculated. Even this range of percentages may not be generous enough.

The remaining yacht weights are conveniently broken down into machinery (if any), equipment, outfit, fuel, stores, crew and baggage. The calculable items such as machinery and equipment should be on the GA and their weights and cgs easily estimated or obtained (remember, we can 'spot' cgs, using judgement). Fuel presents no problems either, because it will be contained in tanks that are either rectangular or else calculated by Simpson's Rule, if a part of the hull side. Items such as outfit, stores, crew and baggage

Table 20.5

Item of weight	Weight Tonnes	VCG m	Vert^L Momt	LCG m	Long^L Aft	Momt Forwd
Hull	1·45	0·60	0·87	0·35A	0·51	—
Deck and Cabin Top	0·30	1·75	0·53	0·20F	—	0·06
Ballast Keel	0·80	0·16	0·13	0·40F	—	0·32
Mast and Rigging	0·10	4·00	0·40	·00F	—	0·10
Engine	0·20	0·90	0·18	1·75A	0·35	—
Accomod – Outfit	0·15	1·25	0·19	0·50F	—	0·08
Shaft and Stern Tube	0·05	0·80	0·04	2·50A	0·03	—
Anchor and Cable	0·08	1·60	0·13	3·50F	—	0·28
Fuel	0·20	1·00	0·20	2·30A	0·46	—
	3·33	0·80	2·67	0·18A	1·45	0·84

$$\text{VCG} = \frac{2·67}{3·33} = \textbf{0·80 m} \text{ Above Baseline}$$

$$\text{LCG} = \frac{(1·45 - 0·84)}{3·33} = \textbf{0·18 m} \text{ Abaft Stat. 5}$$

are less amenable to calculation so they are best estimated using a gross percentage, of say, 10% of the total weight, with the assumption that their combined distribution does not affect the cg position. The various items are then tabulated, as shown in Table 20.5 (and see Fig. 20.6), which, for convenience, is a very much simplified version of a proper weight and cg calculation.

If our weight calculation does not correspond closely with the displacement then we either accept the apparent change of draught, hoping the discrepancy is due to the inevitable approximations involved, or else we change the weight in some way until correspondence is achieved. We can also look more closely at the percentage additions we made for fastenings, connections, etc., and try to make a better estimate by some form of calculation. If the discrepancy is large, then we need to examine the fundamental basis from which we derived the initial value at the preliminary stage. Any incompatibility – whether small or large – should eventually resolve itself without too much difficulty.

A primary condition of equilibrium for any floating body is that its cg lies vertically over the cb. If our calculated cg does not lie over the cb then we either accept the change of trim that will result or else shift weights on board

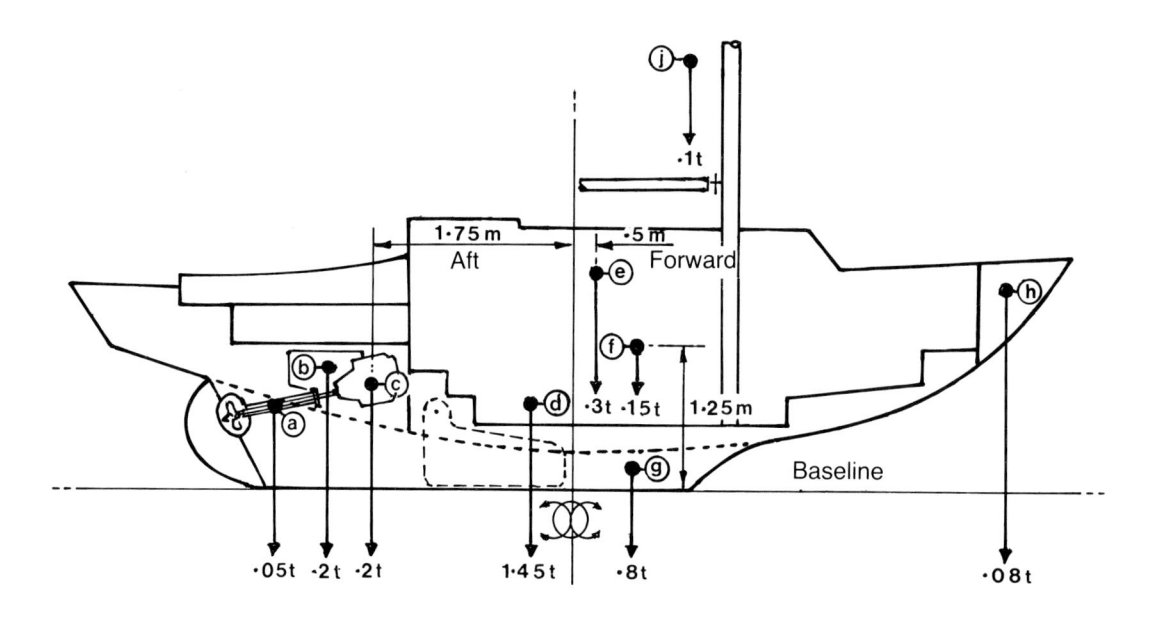

a – Sterntube and Shafting, etc.
b – Fuel.
c – Engine.
d – Hull.
e – Deck and House.
f – Outfit.
g – Ballast Keel.
h – Anchor and Cable.
j – Mast and Rigging.

Fig. 20.6 The cg of yacht is obtained by taking moments of all relevant weights about amidships. A simplified version for such a calculation is shown above.

until the cg and cb lie on the same vertical axis. To find resultant changes of trim, and the weights and moments involved, is a topic more suitable to the stability aspects of a yacht and is therefore included in Chapter 21. The topic covers two cases: (a) the addition or removal of a small weight, which is more suitable to our adjustment of cb and cg in the preliminary design, and (b) the addition or removal of a large weight, which is pertinent to major design changes.

We are now buoyant and aweigh with our design. But will it act like a drunken design at sea, have a heart of oak or timbers that will not shiver? These are the next investigations in our self-design.

21
A DRUNKEN DESIGN AT SEA
The Stability of a Yacht

A yacht can be a glitter of chrome and varnish, with lines suggesting a sea tiger about to spring into foaming speed. Yet no matter how efficient it is in speed and performance, its comfort and safety lie in that part of design known as stability. One tragic instance of this is the sinking of the carrick, *Mary Rose*, mentioned in a previous chapter.

Many similar incidents occurred in the past, explained and unexplained – such as the mysterious loss of vessels without apparent cause during an ocean voyage. Even in present times there is the occasional overturning when circumstances combine to render a ship unstable. Lack of knowledge of the mechanics underlying stability probably caused most of the past tragedies, while nowadays one can design a vessel that is stable under virtually any conditions. Nevertheless, even if we were to make our design ultra-safe, it could be exceedingly uncomfortable to sail in. When heeled over it may spring back like an elastic band – a condition known as *stiffness*. On the other hand, it may roll over and take an eternity to return to the upright position, leaving your stomach behind in the process. This condition is known as *sluggishness*.

I have had the distressing experience of being subjected to both conditions. Trawlers are prone to wallowing and being sluggish in a seaway, due to their shape and type of service. I was once on trials on such a vessel and everything was fine while we were underway; then the order was given to shut off engines. As the boat slowed down, the swell on our port bow began to take over. The hull rolled gently at first, then began swinging over at increasingly greater angles. As my stomach departed from its navel location so, for the first time in my life I sensed that my complexion was matching the colour of the surrounding pea-green sea. Fortunately, before I stained my reputation for never being seasick (and my immediate surroundings), the engine started up again, allowing stomach and pallor to return to normal.

The River Hamble runs into Southampton Water and the English Channel. A few miles from the harbour lay a small boatyard where I once worked, renowned for its unique construction of small craft. It also served as a marina where yacht owners could come for servicing, maintenance and alterations. One day the owner of a very smart motor yacht entered the design office. A self-made man who had built up an electrical engineering business, his overweening confidence led him to claim an infallible expertise in naval architecture.

'I am not happy,' he said, brushing a bristling moustache with his forefinger, 'with the way my yacht rolls. I want you to put a lead weight under the engine to lower the cg.'

Proceeding to instruct me on the elements of my profession, drawing on his threadbare wisdom he outlined a course of design that left me mentally gasping. When inspecting the thin sandwich of space under the engine, I was left even more breathless in the cranium.

'You can't seriously consider getting a slab of lead under *there!*' I protested. 'The cost and trouble won't warrant the advantages – if any.'

As I said, he was a self-made tycoon and no half-baked yacht designer was going to tell *him* what to do. He peremptorily ordered me to 'get on with it' and for days I toiled with the problem of trying to get a pint of lead into a half pint space. Eventually, after much carving round the nuisances of bolts and other projections thought up by the engine manufacturer (which the owner chose to ignore), I managed to insert a hunk of lead into a tight fit under the engine. On trials, the owner's complaint that it had been too sluggish was certainly rectified. There were no long, slow rolling motions – it came back to the upright so quickly I almost knocked my head on the bridge clock. But the owner seemed satisfied; he had a Cheshire-cat grin on his face, so who was I to argue?

What exactly is *meant* by yacht stability?

One very important consideration is the *transverse* aspect, which involves a type of moment known as a *couple*. A coupling moment is one which, as a simple example, turns a knob or dial. The thumb exerts an *upward* force on one side and the first finger a *downward* force on the other side. The distance between them is, naturally, the diameter of the knob or dial and this distance, multiplied by the force acting through thumb or finger, gives the coupling moment. If both thumb and finger were to act in the same direction there would be no rotary motion and hence, no couple. How does this simple concept relate to yacht stability?

Figure 21.1 illustrates three midship sections of a yacht hull, their only difference being in the locations of cg (labelled G in the figure). The buoyancy (labelled B in the figure) remains in the same location, but shifts over to B_1 when the waterline, WL, heels over at a small angle to the horizontal to take up the new position W_1L_1. It can be seen that a triangular wedge is formed either side of the centreline, one being *emerged* and the other *immersed*. In effect, there has been a transfer of buoyancy from one side of the hull to the other, which means that the cb (B) moves in the direction of the immersed wedge. The cg does not move (except if there were a shift of weights on board), so we have a situation where the weight (W) and buoyancy force (both equal) act as a couple with the distance between them being GZ. In Fig. 21.1(a), G is below M, so the couple tends to *right* the hull: a condition known as *positive* stability. In Fig. 21.1(b), G is above M so the couple makes the hull heel further over: a condition known as *negative* stability. Where M and G coincide, as in Fig. 21.1(c), there is no couple acting so the hull remains in the heeled condition, neither returning

A Drunken Design at Sea

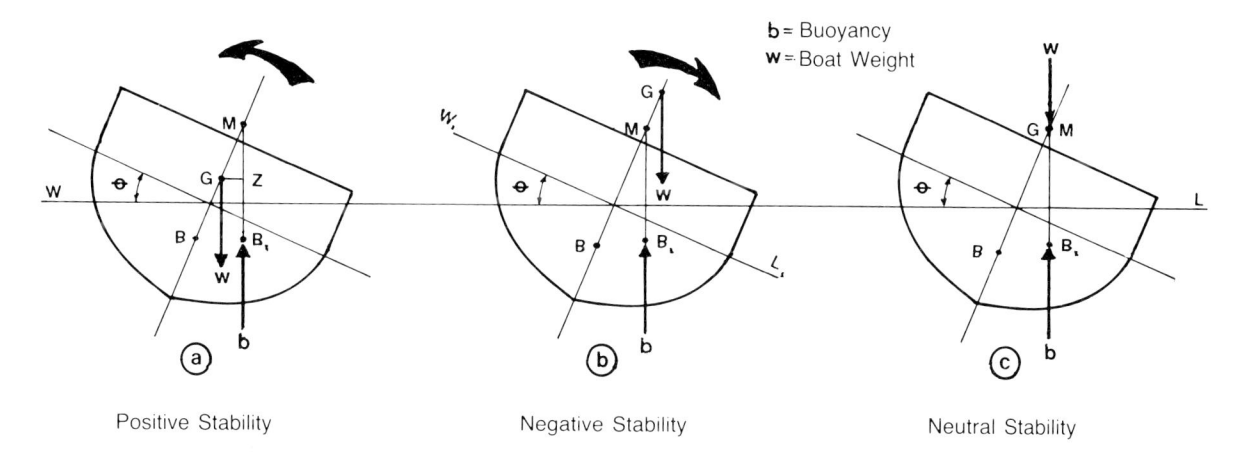

Fig. 21.1 Yacht stability is vital to the safety of the design. The three conditions of transverse stability which are possible when a hull rolls in a seaway are illustrated above, and obviously conditions (b) and (c) are to be avoided.

to the upright nor heeling over further. This condition is known as *neutral stability*. From these three conditions we learn the elementary, but golden, rule that M must always be above G to obtain positive stability and thus, a stable boat. But how *much* above G must M be?

In the three illustrations of Fig. 21.1 it should be noticed that the projected line of the buoyancy force cuts the hull centreline at M (obviously, M and G coincide in the neutral stability condition) and the distance GM, known as the *metacentric height* (the meaning of metacentre is *turning point*), is one which is most important to naval architects and yacht designers. The distance GZ is also very important because it is the governing factor of the righting moment and is therefore known as the *righting lever*. GZ can be calculated using the trigonometrical ratios presented in Chapter 12 and the principle of similar triangles (triangle GMZ is similar to triangles WOW_1 and LOL_1). This triangular configuration is also applicable to pitching and the longitudinal GM, as it is known, is illustrated in Fig. 21.2. Notice that in the longitudinal case the triangles are formed through the centre of flotation (CF) and while pitching is not so critical to design as rolling, the derived formula is almost identical. We can thus show that:

$$GZ = GM \sin \theta \text{ for the transverse condition}$$

$$GZ_L = GM_L \sin \theta_L \text{ for the longitudinal condition}$$

GM for the transverse condition only applies for angles from $0°-15°$. Beyond this the rotation of hull does not occur at the crossing point (0) between the waterline and centreline and therefore, more complicated calculations are required which involve the righting lever GZ. There are a

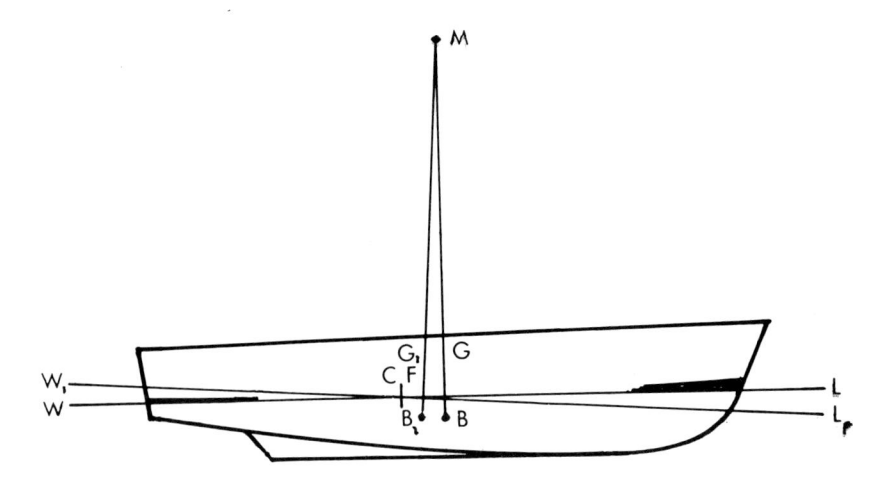

Fig. 21.2 Longitudinal stability (pitching) is similar to transverse stability, where righting levers and metacentric height govern the motions.

number of methods enabling this to be done and some are explained in the final chapter, but for our present DIY purposes, considering the size of craft we are dealing with, it is not essential to conduct a full analysis of stability. All we need do is satisfy ourselves that the initial stability of our hull is within acceptable limits. Simple formulae can be derived to give the height BM for both the transverse and longitudinal condition. Again, they are very similar in nature and are given by:

$$BM_T = \frac{I_T}{V} \text{ for the transverse condition, and}$$

$$BM_L = \frac{I_L}{V} \text{ for the longitudinal condition}$$

I_T and I_L are the moments of inertia of LWL in the transverse and longitudinal directions, respectively, as explained in Table 12.3 of Chapter 12. In order to obtain I_L we need to find the centre of flotation (cf), which is the centre of area of waterplane and calculated as shown in Table 21.1. Adjustment has to be made for I_L to be about cf according to the formula given in Chapter 12. The calculation for the transverse and longitudinal moment of inertia is performed in Tables 22.1 and 22.2 respectively, together with the BMs of the strength chapter (Chapter 22). Knowing the positions of B and M – and the distance BM – it is now quite easy to subtract the distance BG from BM to obtain GM.

So the question now arises – what are acceptable values of GM for initial stability? First, there is a limiting case, as shown in Fig. 21.1(b), though obviously G can be raised before instability occurs. Now, GM is a good guide to stability for angles up to 15° and in general, it can be said that a

A Drunken Design at Sea

Table 21.1

Sta	½ Br	½ Br³	SM	MI funct
0	0	0	1	0
1	0·60	0·216	4	0·864
2	0·90	0·729	2	1·458
3	1·04	1·125	4	4·500
4	1·10	1·331	2	2·662
5	1·07	1·225	4	4·900
6	0·98	0·941	2	1·882
7	0·82	0·551	4	2·204
8	0·58	0·195	2	0·390
9	0·32	0·033	4	0·132
10	0	0	1	0
				18·992

Moment of inertia $= 18 \cdot 992 \times \dfrac{2}{3} \times \dfrac{0 \cdot 65}{3} = \mathbf{2 \cdot 743}$ m⁴

$$BM_T = \frac{I_T}{\nabla} = \frac{2 \cdot 743}{3 \cdot 240} = \mathbf{0.85 \text{ m}}$$

($\nabla = 3 \cdot 24$ m³ taken from Table 20.2)

large metacentric height will produce a quick period of roll while not necessarily contributing to a long range of stability. A minimum GM – desirable in many respects – needs also to provide an adequate range of stability. To solve the problem requires elaborate design procedures and possibly tank experimentation work, not feasible – or necessary – in our case. Indeed, even professional designers limit their stability enquiries, except for large vessels.

To combine minimum GM with an adequate range of stability we can only resort to our basis design, assuming it has good motion characteristics and a satisfactory range of stability, together with guides obtained from past boats. As a selective example, suitable values of GM for displacement type powerboat hulls are about 30 cm (approximately 1'). For other small craft GM ranges between 45 cm and 60 cm (about 1'-6" to 2'). For sailing yachts the metacentric height can be taken as ranging from 0.9 m for boats with a waterline length of 8 m (about 3'-0" for 27-28 footers) to 1 or 1.2 m for 15 m length 3'-3" to 4'-0" for 50-footers). Comparing with the basis design should present no problem as we should know its waterline shape and displacement to obtain BM.

As mentioned previously, beyond, say, 10°–15° we can no longer apply a simple calculation, which means the full range of stability is not easy to

Table 21.2

Sta	½ Br	SM	Area funct	Lever	Momt funct	Lever²	MI funct
0	0	1	0	5	0	25	0
1	0·60	4	2·40	4	9·60	16	38·40
2	0·90	2	1·80	3	5·40	9	16·20
3	1·04	4	4·16	2	8·32	4	16·64
4	1·10	2	2·20	1	2·20	1	2·20
5	1·07	4	4·28	0	0·00	0	0
					2·52		
6	0·98	2	1·96	1	1·96	1	1·96
7	0·82	4	3·28	2	6·56	4	13·12
8	0·58	2	1·16	3	3·48	9	10·44
9	0·32	4	1·28	4	5·12	16	20·48
10	0	1	0	5	0	25	0
			22·52		17·12		119·44

CI = 0·65

Area = 22·52 × 2 × 0·65/3 = **9·76 m²**

CoF = (25·52 − 17·12) × 0·65/22·52 = **0·24 m** Abaft Stat 5

Moment of Inertia = 119·44 × $\frac{2}{3}$ × 0·65³ = **21·87 m⁴**
(about Stat. 5)

Moment of Inertia = 21·87 − (9·76 × 0·24²) = **21·31 m**
(about cf)

$$BM_L = \frac{I_L}{\nabla} = \frac{21·31}{2·93} = 7.27 \text{ m}$$

(∇ = 2·93 m³ taken from Table 20.2)

assess. Nevertheless, it is as well to have knowledge of the *Curve of Statical Stability*, as it is titled, and depicted in Fig. 26.2. For such a curve the righting lever GZ is a better indication of stability over the range and is therefore plotted against a horizontal axis of angle of heel. Some points to note regarding the curve are that for small angles of inclination there is a tangential straight line to the curve which can be drawn at a heeling angle of 57.3° (one radian in trigonometric terms) – see again Fig. 26.2. The maximum GZ represents the largest steady heeling moment and its value and the angle at which it occurs are important; the GZ value eventually reduces to zero (often for angles greater than 90°), becoming negative for larger angles. This angle is known as the *angle of vanishing stability*; and up to this point the boat will return to the upright when the heeling moment is removed.

A Drunken Design at Sea

When the yacht has been built (or if you are fortunate enough to own one already), an exact assessment of GM can be made by carrying out an *inclining experiment*. Two pendulums of known length are placed at suitable distances apart, along and on the centreline of the hull. A length of board is placed behind each pendulum and a mark scribed at the bottom to denote the centreline. On board the deck, at the centreline, will be placed known weights which are then shifted first to one side then the other to obtain small angles of heel (say 3° to 4°). This results in consequent shifts of the pendulums, which are marked on the boards and then averaged. Using the trigonometric ratios, we can obtain the heeling angle through the fact that we know the pendulum length and the shift from the original centreline. This is given by:

$$\tan\theta = \frac{y}{l}$$

where y = average shift and l = length of pendulum. By simple triognometric and moment analysis it is then possible to obtain the following formula:

$$GM = \frac{wd}{\Delta \tan\theta}$$

where w = weights on board and d = distance shifted. For example, if 0.25 tonnes were shifted one metre on a 5-tonne yacht to produce an angle of heel of 4° ($\tan 4° = 0.07$ approx.), then:

$$GM = \frac{0.25 \times 1}{5 \times 0.07} = \frac{0.25}{0.35} = 0.7 \text{ m approx.}$$

Care should be taken before performing the experiment that all liquid tanks on board are pressed up (filled to capacity) for reasons which will be discussed next. Finally, it should be mentioned that it will be necessary to correct for the weights used as they will both sink the boat and raise the CG.

Before introducing some factors that adversely affect stability I would first ask you, the reader, whether you have noticed how much more difficult it is to carry a shallow tray of water than an equal amount in a deeper, narrower container. The water sloshes around in the shallow tray making it awkward to retain a steady hold. This phenomenon is known as *free surface* effect, and adversely affects stability if tanks holding water or any other liquid such as fuel are not pressed up. In such a free surface condition the GM is lowered, or can be considered as a *virtual* rise in cg, this virtual rise to G_1 being obtained by the following formula:

$$GG_1 = \frac{s_f I_f}{s_s \nabla}$$

where s_f and s_s are the densities of the liquid on the tank and either fresh or salt water (depending on which the hull is floating on), respectively, and I_f is the moment of inertia of the free surface. Note that this effect is not at all dependent on the position, volume or depth of water.

Another adverse effect on stability is when there are freely suspended weights. Referring to Fig. 21.3, the weight, w, suspended at s, moves over, and at the same angle to the vertical as the heeling angle. As far as the yacht is aware, the line of action of w passes through s and it is as though the weight were placed at this point.

Having dealt with the major aspects of initial stability, we now turn to the yacht motions in a seaway. This is an extremely complicated subject involving the damping effect of water, the random nature of seas, solution of highly complex equations to take account of the three-dimensional nature of the problems that arise, and so on. Such problems are intensified when one considers that a yacht or boat has six motions – rolling, pitching, yawing, surging, heaving and swaying. The first three motions are rotational (turning) while the last three are either horizontal or vertical movements. Necessarily, only generalities can be dealt with, together with some simple empirical formulae; trying to couple two or three motions for a theoretical solution makes it mathematically prohibitive to solve in our own, and most other, cases.

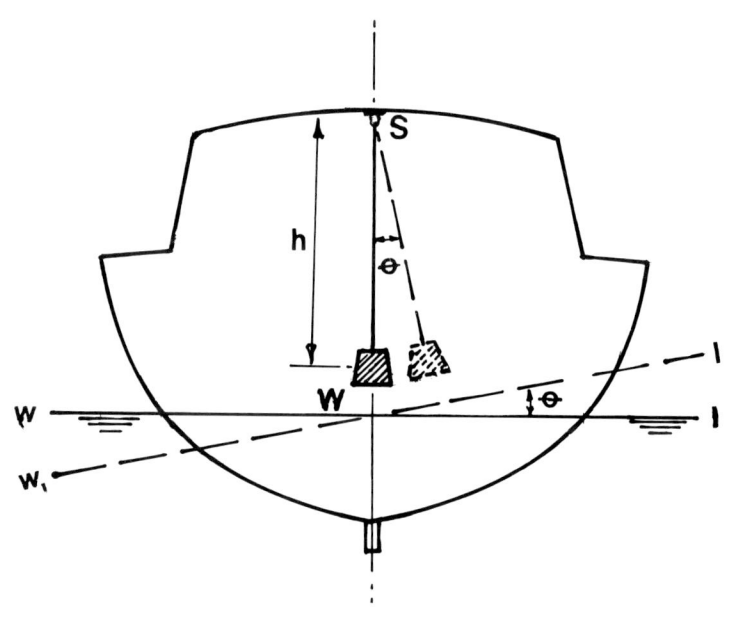

Fig. 21.3 Freely suspended weights as shown above, can adversely affect transverse stability.

A Drunken Design at Sea

In discussing good seakeeping motions of a yacht, excessive amplitudes are undesirable from the point of view of crew and passenger comfort. Every hull has a natural period of roll (assuming it is undamped by the surrounding water) and one approximate formula based on the beam and type of craft is given by:

$$T = C \times \frac{\text{Beam}}{\sqrt{\text{GM}}}$$

where the values of the constant C vary in practice between 0.38 and 0.59. For most vessels, including small boats, and depending on type, C lies in the region 0.4 to 0.45. A best value for C in our case would be derived from the basis design, assuming all other factors relate.

For pitching, the attention is focussed on head seas and one approximate formula is given by:

$$T = \frac{C_p}{L}$$

Values for the constant C_p can vary between 0.22 and 0.5. Probably a suitable figure for most cases is about 0.3 – 0.4, but once again we should be able to derive a C_p for our use from the basis design.

To conclude the chapter we now need to discuss the effect on trim, how to find it and if a small or large weight is added or taken off our design or an existing yacht. A yacht trims about the centre of flotation (cf) and should it float at a different trim to the design trim (say, if we found the cb and cg did not align vertically in our calculation and we were willing to accept the new

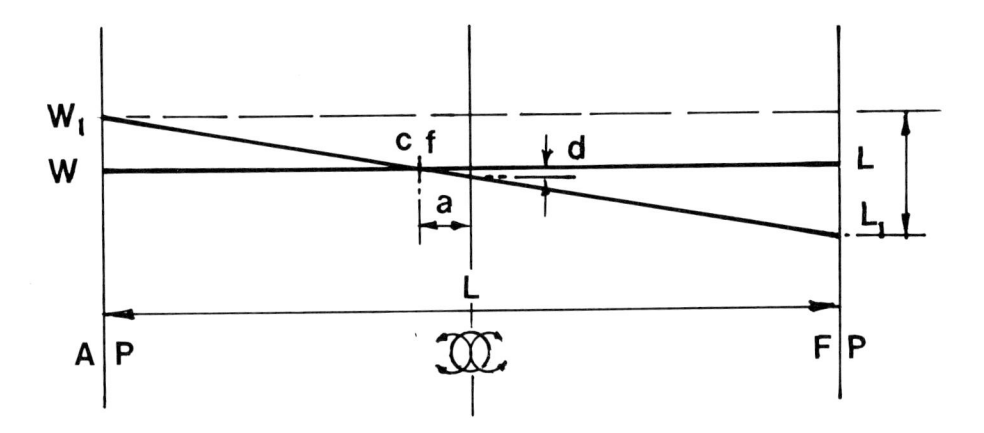

Fig. 21.4 The addition or removal of a small weight will bring about a change of trim and involve a small layer correction to the draught and displacement, as illustrated above.

trim) then using the principle of similar triangles and referring to Fig 21.4, the mean draught correction is given by:

$$d = \frac{at}{\text{LWL}}$$

where a is the distance of cf from amidships, t is the change of trim and d is the change in mean draught. Using this and TPI or TPC, depending on units, we can obtain the trim correction by the following formula:

$$\text{Trim correction} = \frac{at}{\text{LWL}} \times \text{TPI or TPC}$$

Whether it is a positive or negative trim depends on the position of cf and whether the excess trim is by the bow or the stern. In order to establish this it is best to make a sketch similar to Fig. 21.4 and rationalise which is the case for our design.

For a small weight addition (or removal) we assume a weight W is added at a distance h from the cf, then the separate effects of trim and sinkage can readily be calculated by:
(a) Assuming the weight W placed over the cf to cause a parallel sinkage W/TPI.
(b) Calculating the change of trim due to Wh, using Wh/MCT 1″ (or metric equivalent).

For large weight addition or removal the simple assumptions and approximations applied in the case of small weights no longer hold. While it is possible to make a calculation from the existing design, the process is complicated and laborious. A more preferable and direct approach is to regard the yacht as an entirely new design, with revised positions of G and B, together with correct displacement. Naturally, B must again be located vertically under G and the trim determined, and *Bonjean Curves* (see Chapter 25) will be found useful in this respect.

So we now have a few stability facts at our mental fingertips; the next progression of design is to investigate the strength requirements of our design.

22

HEARTS OF OAK

The Strength of a Yacht

Listen to a Royal Marine band playing at a naval function and inevitably at some time during the programme it will strike up that most traditional of seafaring tunes, so closely associated with the British Navy – *Hearts of Oak*:

> *Hearts of oak are our ships*
> *Hearts of oak have our men . . .*

The wooden ships-of-the-line were framed with oaks from the great forests, specially grown to supply the shipyards of a navy that truly ruled the waves in Nelson's time. The vessels were sturdy and able to withstand the broadsides of French and Spanish men-of-war. The Royal Navy's dominance of the seas lay in the timbers and craftsmanship that went into them, as well as the highly disciplined men and officers. Thus, the rousing tune and words embody the feats of seamen and shipwright, whose union brought about the finest period in British maritime history.

The modern yacht or motor cruiser is more likely to be built of GRP, alloy or steel than wood, but the constructional principles still apply – including the assurance that they are designed strong enough to resist the poundings of wave and weather.

During the dark days of the Second World War the British Expeditionary Force in France suffered many defeats – and a long retreat. British morale was then boosted by probably the largest rescue operation of an army ever mounted. A flotilla of small craft and naval vessels sailed one morning from the southern shores of the island to lift off the thousands of British and French soldiers stranded – and under continuous attack – on the Dunkirk beaches. In all, nearly 340,000 troops were brought back and many of the small craft were abandoned up river or estuary after they had completed their mission, their condition not worth salvaging.

The media spotlight focussed on the Royal Navy's exploits after this spectacular retrenchment; the Royal Navy – bastion of defence for so many centuries – now guarding pipelines of sea trade, transported in convoy fashion.

I remember seeing the newsreels at the time – as well as a number of fictionalised, but patriotic films of the Navy's sea exploits. Standard Saturday afternoon cinema fare (we were not allowed to go in the evening because of the air raids) was (along with westerns and gangster films) wartime sea sagas such as *In Which We Serve* and newsreels showing small

Design Your Own Yacht

naval vessels guarding the convoys to Russia. As a wide-eyed youngster, it always amazed me how those flimsy frigates and corvettes survived the batterings of nature, burying their bows into mountainous seas and yet emerging safely with a huge spill of water. Even now, though equipped with the technical know-how, I still feel momentary wonderment.

I'm quite sure that many readers have had slight feelings of apprehension before crossing what appears to be a fragile suspension bridge. In fact, such delicate structures are designed with a lattice of steelwork to withstand immense loads. A yacht (and ship) is designed on similar principles to a bridge or beam, and to resist equally strong forces. You may well ask how this is possible if it is floating on water. The answer is: waves.

Let us take two extreme examples, one where the yacht is suspended between the crests of a wave equal to its own length, and then on just one crest at amidships. In the first case (Fig. 22.1(a)) the hull will tend to sag and this is known as the *sagging* condition. In the second (Fig. 22.1(b)) it tends to hog and this is therefore known as the *hogging* condition. Such conditions obviously set up extreme stresses at either amidships or near the ends. Now, stresses are caused by forces; in this case, the resultant forces due to the buoyancy forces acting upwards and the weight of hull and outfit acting downwards. It must be borne in mind that for very small craft slamming loads tend to dominate these loads.

To analyse the stresses acting on the hull we must look to that area of structural engineering known as *Strength of Materials*. As always in this book, I will use a simple analogy to depict the situation. Imagine, if you will, a matchbox being bent under the pressure of fingers (Fig. 22.2). In part (a)

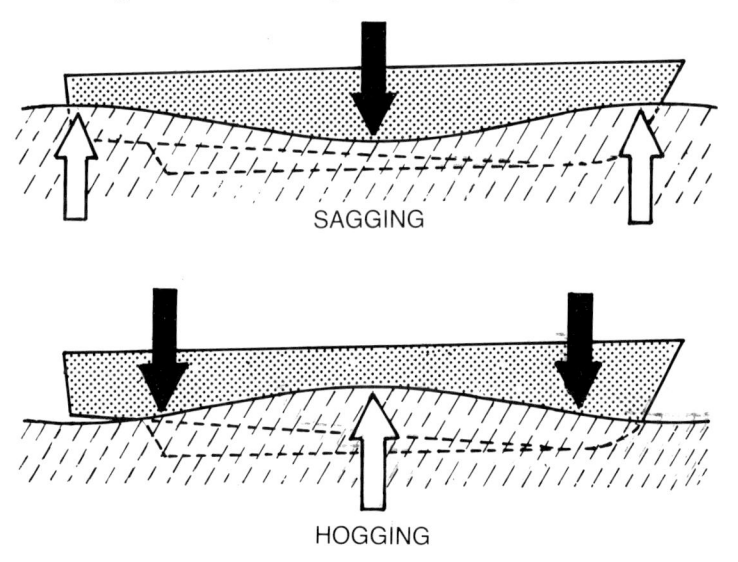

SAGGING

HOGGING

Fig. 22.1 A yacht hull is like a floating bridge, especially when on the crests of two waves such as shown in Fig. (a), above, which is called the sagging condition. Figure (b) illustrates a hogging condition of hull.

Hearts of Oak

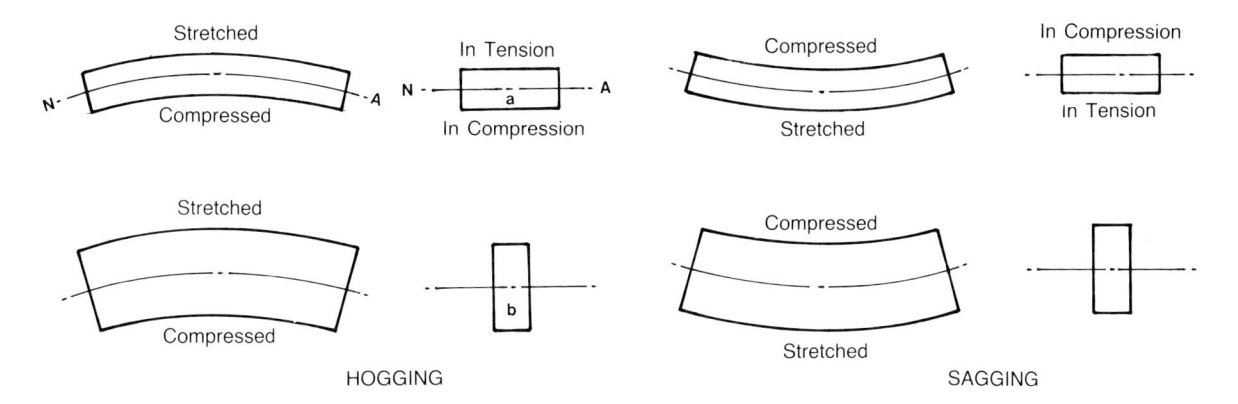

Fig. 22.2 When bending a matchbox it will be quickly realised that the alignment of cross-section to the pressure applied is important. The alignment in Fig. (a) makes it easier to bend than in (b), and it is this materials property that is considered in calculating the strength of yacht hulls.

of the figure the matchbox is easier to bend because its width is horizontal; when the width is placed vertical – as in part (b) – the matchbox requires much greater force to bend it even a little.

When analysing the situation, what seems obvious by experience proves to be correct by mathematical considerations: the *depth* of a cross-section governs the resistance to bending.

We now have to look at what type of stresses are acting, and the point of greatest intensity. For a symmetrical shape such as a matchbox there is a plane where no stress is acting and this is at its centre of area. This plane is known as the neutral axis. If the fingers are bending the matchbox downwards (the hogging condition) then above the neutral axis the matchbox material is being *stretched*, to produce *tensile* stresses, while below it is being *compressed*, to produce *compressive* stresses. Obviously, the reverse or sagging condition will produce tensile stresses in the bottom and compressive stresses at the top.

The greatest stresses will occur at the extremes of a cross-section, and that is why structural girders and beams are all long, narrow shapes such as channel bars, tee-bars, box girders, etc. (Fig. 22.3). The *flanges* at the ends, as they are called, are also much thicker than the vertical parts of the section in order to produce an efficient section resistant to tensile and compressive stresses. Later, we will present the formula that determines these stresses, but for the moment it may be said that such stresses are entirely dependent on the loads acting on a structure. In our case we have to find the differences between the distribution of the buoyancy forces and the weight forces and this is done by means of diagrams. The calculation for total hull weight has already been described in Chapter 20 and it only requires unit strips of weight to be calculated to produce a weight curve for a hull. There are also

Fig. 22.3 Commonly used sections in structural and marine engineering.

approximate methods that can be used and one technique to approximate hull weight is presented in Fig. 22.4. Upon the hull weight curve are then drawn the other yacht weights such as engine, mast, and other significant loads that may be presented as rectangles over short distances. More ambiguous loads such as outfit, rigging, etc., can be considered as being distributed over the hull length and the total weight curve is as depicted in Fig. 22.5(a).

The next diagram, Fig. 22.5(b), deals with buoyancy. The hull is placed on a wave equal to its own length, depending on the condition being calculated (hogging or sagging), and using Simpson's Rule, the buoyancy forces are calculated for unit strips to produce the diagram as shown in the figure. It is not easy to draw this curve to suit the displacement and cb and a more detailed explanation is given in Chapter 26. The difference between this curve and the weight diagram will produce the Load Diagram shown in Fig. 22.5(c).

The summation of loads or forces acting from a certain datum line to a point X – as depicted in Fig. 22.5(c) – will produce the Shearing Force Diagram as illustrated in Fig. 22.5(d). This diagram shows the shearing

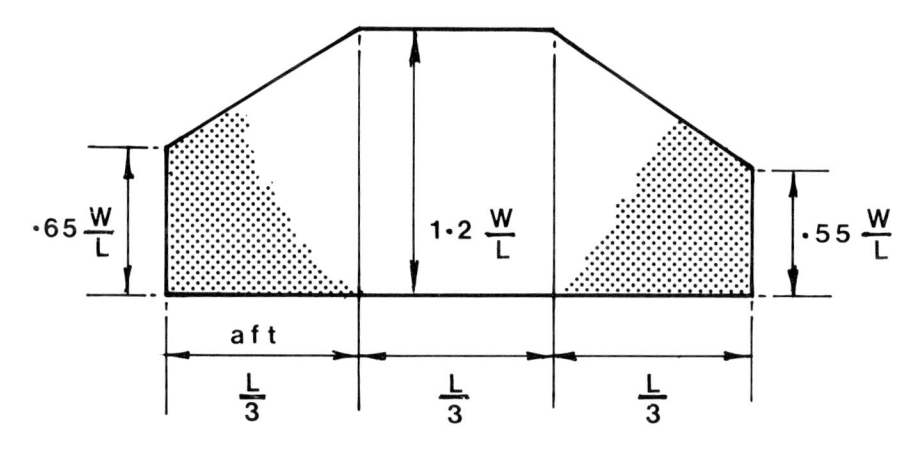

Fig. 22.4 The weight curve for a hull can be approximated by using a diagram such as above.

forces acting on the hull, which are equal at each point to the area of the Load Curve up to that point (shown shaded in the diagram).

From this diagram (calculated for both the hogging and sagging condition) is produced the diagram of the Bending Moment (Fig. 22.5(e)) acting on the yacht. To calculate this curve we find the area under the Shearing Force Curve up to a point such as X, which then gives the BM at that point. It is from these diagrams that calculations are made to determine the size of scantlings such as hull thickness, frames, beams, etc. Nevertheless, the work involved is quite considerable and one approximate formula for maximum bending moment at the early design stage is given by:

$$M = \frac{\text{LWL} \times \nabla}{35}$$ (for a standard L/20 wave of the hogging condition)

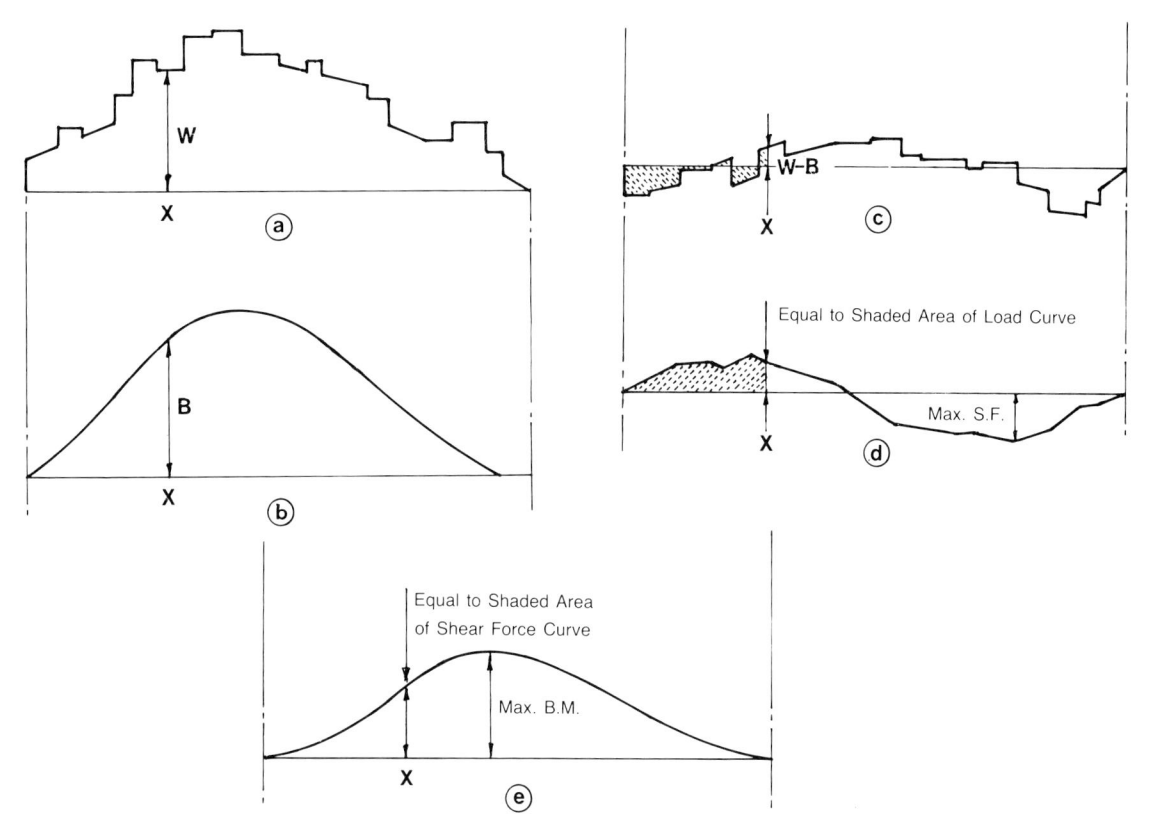

Fig. 22.5 The forces and stresses acting on a yacht hull are obtained by means of diagrams and curves such as the simplified versions above. Figure (a) is a Total Weight Curve for yacht and (b) the Buoyancy Curve on two crests of a wave. Figure (c) is the resultant Load Diagram obtained from the other two curves, while (d) and (e) are the Shearing Force and Bending Moment diagrams, respectively, derived from the others.

where the displacement is in tons and LWL in feet. It must be pointed out that values will be somewhat high for fine-ended hulls and low for full-ended ones. Discretion should therefore be applied when using this formula, while for shallow draught, light displacement types, even greater caution is required.

Now, maximum shearing forces usually occur at the quarter length positions of a hull, and one approximation is to assume about one-seventh or one-eighth of the displacement. Having determined the maximum bending moment and shearing force by whatever means, the stresses acting are then calculated using the following formulae:

$$f = \frac{MY}{I} \text{ (for bending stress)}$$

$$q = \frac{FA\tilde{Y}}{Ib} \text{ (for shear stress)}$$

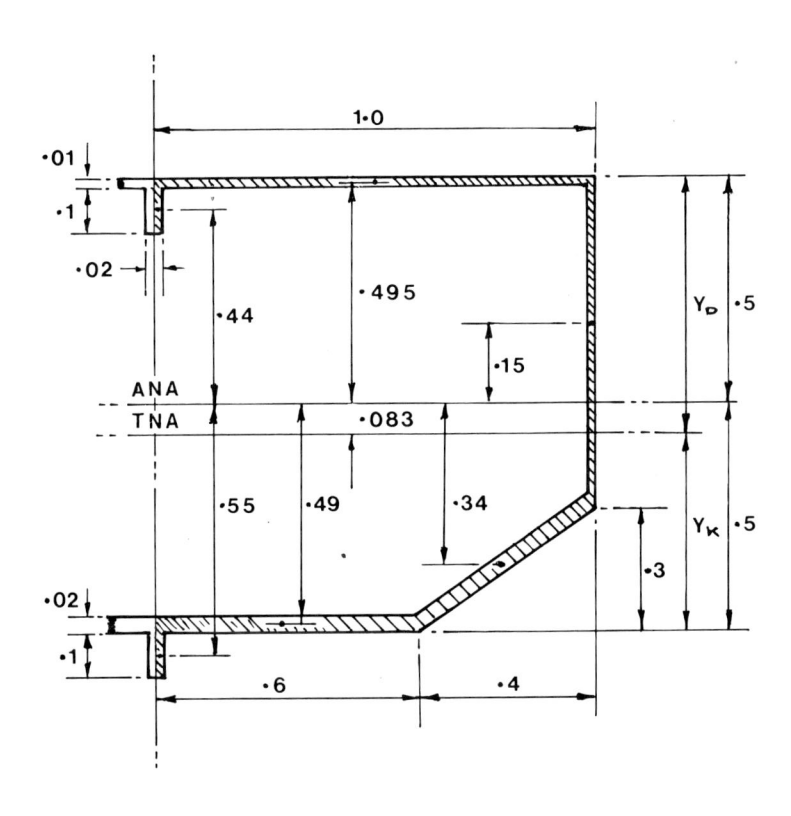

Fig. 22.6 The above simplified Midship Section illustrates how the maximum stresses acting are determined by calculating the Moment of Inertia.

Table 22.1

		Col 1	Col 2	Col 3	Col 4	Col 5	Col 6	Col 7	Col 8	Col 9	Col 10
Below ANA		Keel	10×1	10	·55	5·5	·303	3·03	·10	·01	·1
		Bottom Shell	60×2	120	·49	58·8	·240	28·80	·02	—	—
		Bilge Shell	50×2	100	·34	34·0	·116	11·60	·30	·09	9·0
						98·3					
Above ANA		Side Shell	70×1	70	·15	10·5	·023	1·58	·70	·49	34·3
		Deck	100×1	100	·495	49·5	·245	24·50	·01	—	—
		Deck Girder	10×1	10	·44	4·4	·194	1·94	·10	·01	·1
				410		64·4		71·45			43·5

Where Col 1 Item of Structure   Cm
            2 Size of Item   Cm
            3 Area (one Side only) Cm^2
            4 Distance of CG from ANA   M
            5 Moment about ANA (Col 3 × Col 4)   Cm M
            6 Distance from ANA^2 (Col 4 × Col 4) M^2
            7 Col 3 × Col 6 $Cm^2\ M^2$
            8 Effective Depth of Item   M
            9 Col 8 × Col 8 M^2
           10 Col 9 × Col 3 $Cm^2\ M^2$

True Neutral Axis = Excess of Moment Above or Below ANA ÷ Total Area

$$= \frac{Col\ 5}{Col\ 3} = \frac{(98·3 - 64·4)}{410} = \textbf{0·083 m Below ANA}$$

Moment of Inertia = $2\ (Col\ 7 + \frac{Col\ 10}{12} - Col\ 3 \times 0·083^2)$

$$= 2\ (71·45 + \frac{43·5}{12} - 410 \times 0·083^2)$$

$$= 2\ (71·45 + 3·62 - 2·82) = \textbf{144·5}\ cm^2\ m^2\ \text{or}\ \textbf{1445000 cm}^4$$

Section Modulus

At Keel = $\frac{1445000}{(50 - 8·3 + 10)}$ = **27950** cm^3

At Deck = $\frac{1445000}{(50 + 8·3)}$ = **24786** cm^3

where q = the shear stress, f = the bending stress, M = the maximum bending moment, Y = the distance from the true neutral axis (TNA) to the deck or keel (Y_d or Y_k in Fig. 22.6), F = the maximum shearing force acting, $\bar{Y}$ = distance to centre of Area A being considered, above or below TNA, and b = the total width of plating at that section. The I and other related values for shearing stresses are taken where the maximum shearing forces occur, while the I and Y for maximum bending stresses are taken at amidships and I/Y, the moment of resistance corresponding to unit stress, is termed the *modulus of the section*. The calculated values for f and q must then allow for safety factors so the results, and ultimate sections obtained, are safe working stresses. Many structural engineering handbooks will provide this information against suitable I/Y values for different sections (bending stress) or similar for shear stress requirements.

It still remains to make a calculation for I, whatever the hull section being considered, and for this we must assume a section and sizes that will provide the safe working stresses necessary for the yacht to have a heart of oak. The basis design may provide useful guidance in this direction and once we have drawn a section that seems reasonable (Fig. 22.6 is a simplified version to ease the calculation and relevant explanation) then we perform the calculation to obtain I as shown in Table 22.1. When put through the formula, should the result not comply with appropriate standards, then the sections must be 'beefed' up until suitable values are obtained.

A complete analysis of stress would require a combination of the bending and shearing stresses, too complicated to enter into here and not necessary for our purposes, and in fact it is true to say that all one needs do is calculate the bending stress for deck and keel at amidships and the shear stress at the neutral axis (where the maximum occurs) at the quarter ends. Even more simple would be to obtain guidance from a Classification Society's Rules and Regulations such as that of Lloyd's Register of Shipping. This is discussed in the next chapter, when we make sure that not only does our yacht design have a heart of oak, but also that its timbers do not shiver.

23

TIMBERS THAT DO NOT SHIVER
Construction Plans

In the past, construction of a sailing ship was very much a matter of experience gained over the centuries and becoming familiar with the available materials of construction – mainly wood. As time progressed so this experience was enhanced by knowledge gained on strength properties, best methods of connection and the theory that emerged in the course of building bridges and other structures. This all combined to make a hull a strong, floating girder.

Looking in general at the major constructional design features of a yacht or boat hull, the longitudinal members such as deck and keel longitudinals provide resistance to the bending stresses induced by hogging or sagging, and should be continued without interruption (even piercing bulkheads) for as far as possible. A recommended value is about one-third the boat's length either side of amidships. The transverse members such as frames, floors and bulkheads not only provide a framework for the skin of the hull, they also resist the transverse and racking stresses. Bulkheads in particular are essentially strong elements of the transverse strength, as well as providing watertight subdivision for large yachts and boats.

On a lesser level, in order to save material and weight, the thickness of hull will taper off from amidships to a suitable value at the ends, depending on the strength calculation or Classification Society requirements, if the yacht is being designed under such rules. This will usually not apply for very small yachts as the constructional inconvenience involved is not worth any weight or cost advantages gained. Regarding the framework connections, in general they should not be abrupt but 'smoothed' by means of a triangular bracket or similar. Obviously, GRP hulls, being homogeneous, will not have such complicated connections as separate brackets and knees, but nevertheless the principle of distributing the stress at corners still applies. Similar reasoning prevails when making openings in bulkheads and deck, such as doorways, hatchways and so on. Their corners should always be rounded as there is a dangerous concentration of stress in way of square corners. Metal hulls will have circular or elliptical holes cut in way of floors and other large, flat areas of plate. These lightening holes, as they are called, not only save weight without any great loss of strength, but have other advantages such as better access and ventilation.

We now come to the drawings that, metaphorically speaking, will not make the timbers of our hull shiver. There are a number of such plans to be drawn, the major ones being the Midship Section, Construction Plan, Bulkhead Plan, Engine Seating Plan (if the yacht has an inboard engine), Deck Plan and Superstructure Drawing. Some of these may be merged in one drawing; for instance, the Midship Section, Construction Plan and Deck Plan could all be incorporated in one large drawing. From these major drawings will generate any other detailed plans according to the individual design, to make a constructive whole of the design.

I spent many years as a junior designer drawing detailed construction plans such as mountings for pumps or brackets for structural parts of hulls. I also drew the gamut of major construction plans such as the Midship Section and my experience is that they *can* be satisfying to draw – provided you have the relevant information to hand and guidance on strength aspects and connections. It will help if you can obtain similar plans from the basis design as they will provide assistance and assurance in your drawing endeavours.

Many of my first drawing tasks were for steel hulls of small craft and I had the backing of a large stock of plans in the office. Nevertheless, I was always apprehensive when it came to printing instructions on the plans, especially those for welded connections, which can involve such tricky areas as plate preparation, type of weld, and throat thickness. There are other areas where the yard worker will have more knowledge than your printed instructions dictate.

I recall once providing long-winded and extremely detailed directives on a construction drawing – the result of much research and assimilation of handbook data – then having the yard foreman bear down on me with a print of the drawing in his hand. In the rich language only yacht and shipbuilders have cultivated, he proceeded to bring me down to design reality, pointing out that most of the instructions were impractical to implement with the existing yard equipment, and anyway did not conform with yard practice. Lastly, he suggested, with four-letter word adjectives, that it would require a shipwright with an English degree to interpret my literary efforts on the plan. Since then I have usually consulted the yard as to drawing requirements or else kept instructions as brief as possible.

The cornerstones of all yacht and boat construction drawings are the Midship Section and Construction Plan. The Midship Section shows the sizes of frames, floors, and longitudinals in that region, as well as thickness of deck, keel and hull. These sizes and thicknesses are obtained either through direct calculation as explained in Chapter 22, or else from the rules and regulations of a Classification Society such as Lloyd's Register of Shipping. Drawing the plan is quite simple; all that is required is a view of the midship contour drawn to a large scale, with thick outlines to denote the hull thickness. To this is added the relevant frames, floors, flats and longitudinals in way of the section. A side or elevation view may also help to show longitudinal members in part. For GRP hulls, because they are homogeneous the structure will be quite simple, with built-in frames and other structural members, but for wood, steel or alloy the frame, floor and other members together with their bracket connections need to be shown and defined on the drawing by their sizes. Hull, deck and keel thicknesses must also be indicated. A simplified typical Midship Section for metal hulls was illustrated previously in Fig. 2.4, showing relevant connections, etc. Plastic hulls, obviously, have homogeneous connections.

The beginning format for the Construction Plan follows that of the GA and Lines Plan, with an Elevation, Plan View and a number of sections on the right that typify sections along the hull. One of these will be the Midship Section, while the plan should represent the deck. Another Plan View could be drawn to show the inner structure at the bottom of the hull. One point to make: the Elevation is usually drawn at the centreline of the ship, while other part elevation views may be shown in way of main longitudinal members. We already appreciate that frames have to be, indicated, as seen from our other drawings, and then bulkheads, floors and flats are also drawn in the first instance. These initial steps are shown in profile for a wooden yacht, as seen in Fig. 23.1. Later development will include more detailed aspects such as connections, thicknesses along the hull and variations in frame and floor sizes along the length, necessary to the total constructional side of design.

The Bulkhead Plan will show all the major bulkheads, together with their stiffening members, usually vertical, although there will be some horizontal stiffeners at relevant points. A full-face or transverse contour of a bulkhead is drawn, as well as a side view or Elevation to indicate longitudinals and their connections.

The Deck Plan can be incorporated as part of the Construction Plan and shows deck beams, longitudinals, hatch openings and the outline of cabins and superstructure, as well as beam–frame connections, beam knees and other structural parts such as doubling plates, hatch coamings, etc. The underneath structure will be shown dotted in the plan, while elevation views will show relevant deck longitudinals and connections to bulkheads.

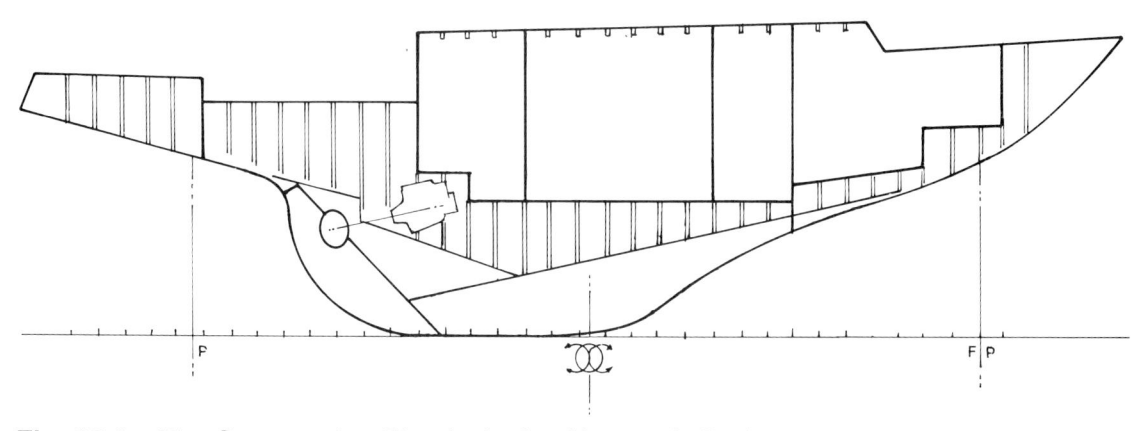

Fig. 23.1 The Construction Plan is the backbone of all other structural drawings in yacht design. Initial steps are to draw the contours of hull, frames, bulkheads, flats, etc., as shown in this profile for a wood yacht.

Plate 26 The Midship section and other sections through the yacht need to be drawn to show the size and location of scantlings along the hull.

If the yacht has an inboard engine then an Engine Seating or Mounting Plan has to be drawn. This will show the main bearer supports for the engine, along with bracket stiffeners and other structural connections to the end bulkheads. The nature of the structure will depend on engine type, its mountings and alignment of shaft. For inboard/outboard or merely outboard engines, the structural reinforcement at the transom needs to be drawn, again dependent on similar factors to those for inboard engines.

There are a number of other plans that have to be drawn: plans of superstructure, fuel and water tanks, if any, flats, chain locker, plus other structural reinforcements that may occur in way of heavy loads such as the mast. It is impossible to list the various detailed plans of construction, but rest assured that drawing the major plans will provide the insight and skill to follow through with such other plans.

Having outlined the constructional requirements of design, we now need to refer to the Classification Society approach in obtaining scantlings. If your yacht design is to come under such rules then it is essential that you have knowledge of their make-up and how they are applied.

Lloyd's Register of Shipping (probably the largest society in the world that deals in the classification of ships and yachts) is selected to show how one applies such rules to a design. Other well-known Classification Societies are the American Bureau of Shipping (ABS), Bureau Véritas of France, Norske Veritas of Norway and Germanischer Lloyd of Germany, but the general approach is the same.

The manual by Lloyd's Register of Shipping is called *Rules and Regulations for the Classification of Yachts and Small Craft*. It covers every aspect of design and construction, from hull to materials and equipment. It

Timbers that do not Shiver

Plate 27 A number of other structural plans need to be drawn to show the structure, scantlings and various connections of a yacht.

is only possible to briefly survey these rules so the reader is aware of the contents, and give selective examples of how to determine the size of certain scantlings.

The contents of the manual are divided into three parts: regulations, hull construction, and machinery and electrical installations. Part 1 of the rules gives general information on classification and surveys, while part 2 deals with hull construction, of interest to this chapter. This section describes the application of the rules, which apply to sea-going motor, sailing and auxiliary craft of normal form and proportions, not exceeding a scantling length of 50 metres (about 165 feet). Craft of unusual design, form or proportions are considered individually. Builders have to demonstrate to the satisfaction of the Society that they have the organisation and capability of building the proposed craft to the standards of the Rules. Plans to be submitted are the GA, Hull Construction Profile, Hull Construction Sections and Bulkheads, Deck, Integral Oil Fuel and Water Tanks, Machinery Seatings, Superstructure, Deckhouses and Coachroofs, Rudder and Steering Arrangements and Propeller Brackets.

Part 2 of the Rules covers the hull construction of GRP craft, steel and aluminium alloy, wood and composite yachts and small craft. All sections follow similar requirements; for instance, in this part of the Rules are given definitions for length, breadth and depth, all measured in metres, while speed, V, is the maximum speed in knots. Quoting from the Rules, the overall length (L_{OA}) is the distance measured parallel to the static load waterline from the fore side of stem to the aft side of stern or transom, excluding rubbing strakes or other projections. The waterline length (L_{WL}) is the distance measured on the static load waterline from the fore side of stem to the aft side of stern or transom. Amidships is to be taken as the middle of the static load waterline, while breadth (B) is the extreme breadth measured between the outer sides of hull, excluding rubbing strakes or other projections. Depth (D) is the distance at amidships, measured from the bottom of keel, or ballast keel if fitted, to the top of the upper deck or gunwale at side. The scantling length (L) is to be taken as:

$$L = \frac{L_{OA} + L_{WL}}{2} \text{ metres}$$

The Rules for hull construction of GRP, steel and aluminium alloy, wood and composite follow similar requirements throughout the manual. For instance, the GRP section includes materials, construction process, workshop requirements, determination of scantlings, hull laminate, internal hull structure, deck and superstructure. These Rules are based on the use of an unsaturated polyester resin system with glass reinforcement, and also on the production of craft by hand lay-up contact moulding techniques and on the use of either single-skin or sandwich construction, or a combination of both. In each case, if there is a departure from requirements, details must be submitted. For steel and aluminium alloy craft the material must be of a marine grade, and the Rules general welding requirements must be based on the electric arc process. The wood and composite section defines the timber species which may be used for various constructional members, together with timber quality, moisture content, plywood and a table for guidance on the selection of timbers for constructional members.

The determination of scantlings – a core requirement of our design if it is to come under Classification Society Rules – is based on a close inspection of such Rules and associated tables. The Rules have formulae and guidelines that may need to be applied to your design and that is why a copy of the Rules is vital in order to comply with Classification requirements. The frame spacing for wood and composite yachts is given in table form, with adjustments or modifications to be made to the section modulus if the spacing differs; the reference frame spacing for GRP, steel and aluminium alloy recommended by the Rules to be adopted in the design is about ($350 + 5L$) mm.

The Rules illustrate requirements for various structural aspects such as keel reinforcement and connections. One such diagram from the Rules is illustrated in Fig. 23.2. For wood and composite yachts, illustrations of terms used in the Rules are also given, and these are shown in Figs. 23.3 and 23.4. The tables, formulae and guidelines in the Rules cover every aspect of hull structure, including transverse or longitudinal framing, bulkheads, stiffeners, rudder, propeller brackets and so on. A selective sample of tables – unfortunately without guidance notes due to space limitations – are

Rule width of keel

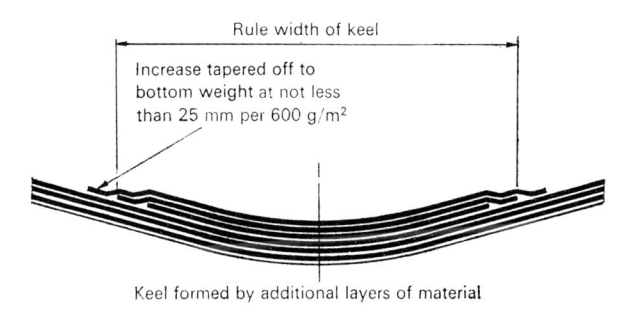

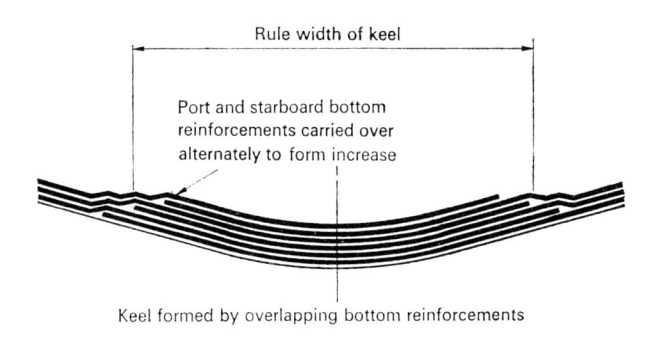

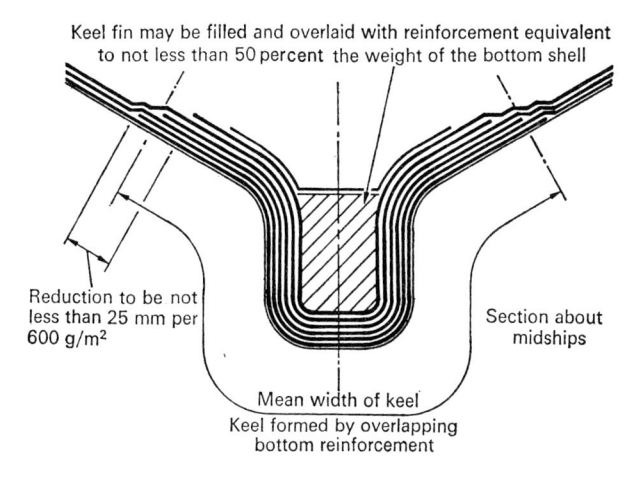

Increase tapered off to bottom weight at not less than 25 mm per 600 g/m²

Keel formed by additional layers of material

Rule width of keel

Port and starboard bottom reinforcements carried over alternately to form increase

Keel formed by overlapping bottom reinforcements

Keel fin may be filled and overlaid with reinforcement equivalent to not less than 50 percent the weight of the bottom shell

Reduction to be not less than 25 mm per 600 g/m²

Section about midships

Mean width of keel
Keel formed by overlapping bottom reinforcement

Fig. 23.2 Lloyd's Rules for yachts and small craft illustrate structural aspects of yachts and small craft, together with formulae and tables. The above illustration from the Rules shows keel reinforcement for GRP hulls. (*Courtesy Lloyd's Register of Shipping.*)

1. Ballast keel
2. Wood keel
3. False keel
4. Lower stem or forefoot
5. Upper stem
6. Sternpost
7. Deadwood
8. Counter timber
9. Rudder
10. Rudder stock
11. Rudder heel fitting
12. Frames or timbers
13. Strap floors
14. Wood floors
15. Beam shelf
16. Clamp
17. Bilge stringer
18. Mast step
19. Outside planking
20. Bulwark
21. Transom
22. Strong beams
23. Ordinary beams
24. Half beams
25. Carling
26. Tie rods
27. Lodging knees
28. Hanging knees
29. Under deck chocks
30. Coachroof coaming
31. Deck
32. Margin plank
33. Covering board
34. Kingplank
35. Coachroof beams
36. Mast partner chock
37. Coachroof top
38. Keel bolts

Fig. 23.3 Every aspect of structure and construction is covered by Lloyd's Rules, whose illustration above gives the terms used in a typical Construction Profile and Deck of an auxiliary craft. (*Courtesy Lloyd's Register of Shipping*.)

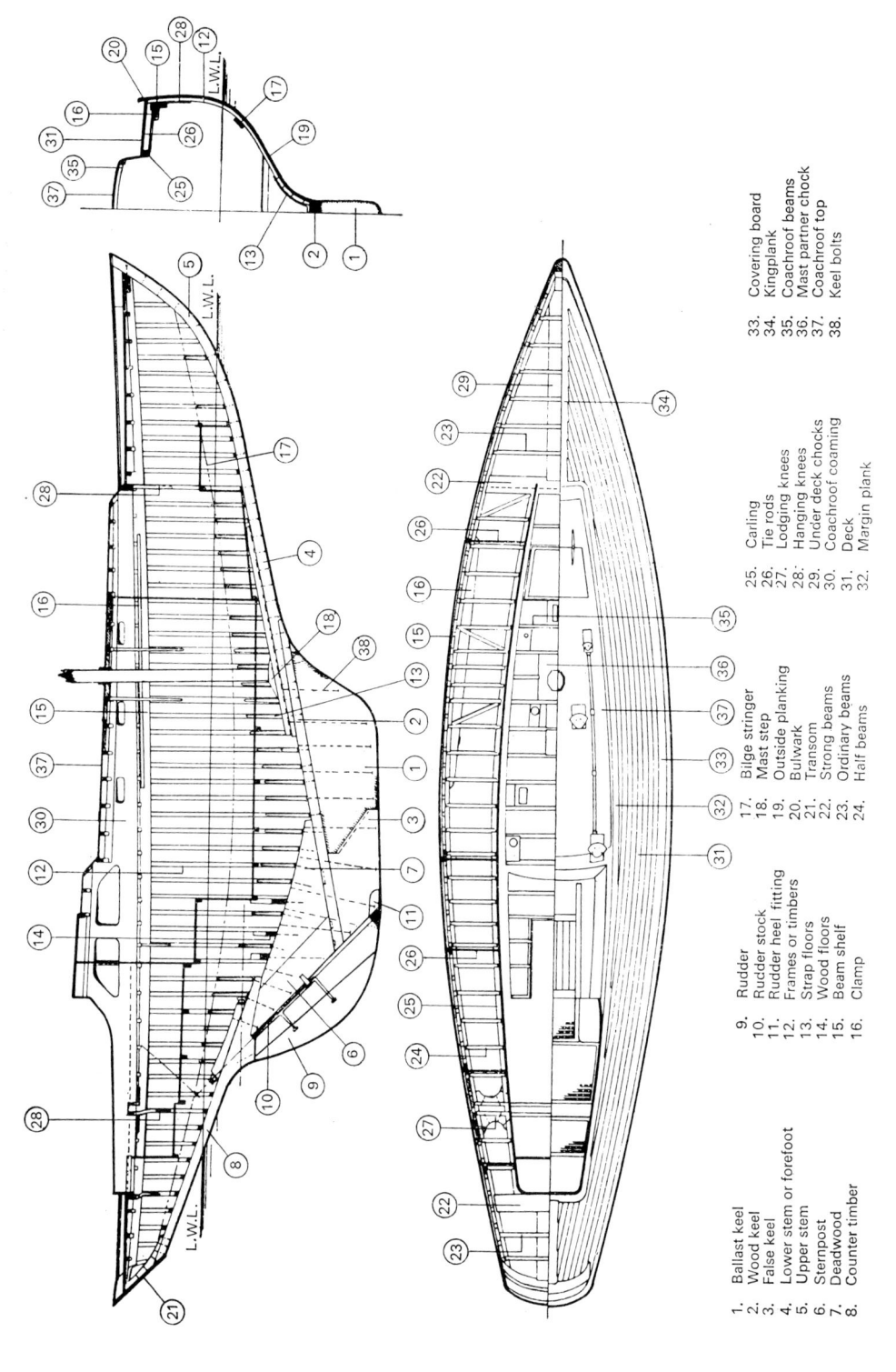

Timbers that do not Shiver

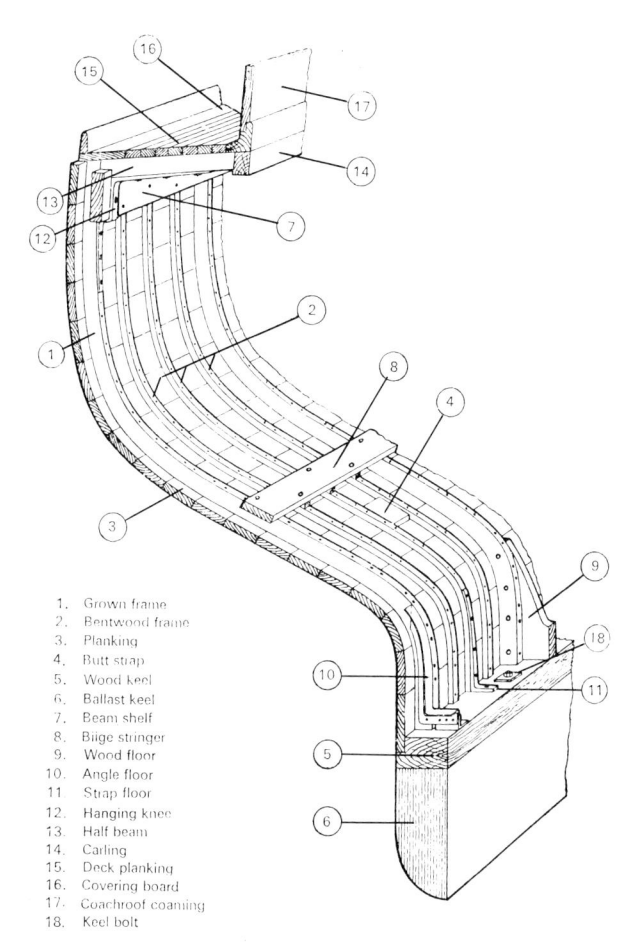

1. Grown frame
2. Bentwood frame
3. Planking
4. Butt strap
5. Wood keel
6. Ballast keel
7. Beam shelf
8. Bilge stringer
9. Wood floor
10. Angle floor
11. Strap floor
12. Hanging knee
13. Half beam
14. Carling
15. Deck planking
16. Covering board
17. Coachroof coaming
18. Keel bolt

Fig. 23.4 An important element of yacht design plans is the Midship Section. The above Midship Section from Lloyd's Yacht Rules illustrates the terms for the Midship Section of a sailing or auxiliary craft. (*Courtesy Lloyd's Register of Shipping.*)

shown in Fig. 23.5. These tables are shown for illustrative purposes and should not be used out of context from the manual.

Other sections of the Rules cover such aspects as safety arrangements, fire protection and equipment, while the latter part covers machinery and electrical installations, as well as control engineering systems – not the direct concern of this chapter.

So we have now covered the constructional aspects of design and are nearing the end of our creative journey. The final steps are to take us into the ancillary areas such as ventilation, lighting, and the information necessary to back up our previous design chapters, such as useful data and tables. It is the slop chest of our design, if you like, which we now have to delve into.

Type 1 — Bent wood frames only

Depth, D, m (Motor)	Depth, D, m (Sailing and auxiliary)	Siding, mm	Moulding, mm	Frame spacing, mm
1,5	1,8	24	19	155
1,8	2,1	34	25	170
2,1	2,4	40	30	185
2,4	2,7	48	36	200
2,7	3,0	56	40	215
3,0	3,3	65	45	230
3,3	3,6	—	—	—
3,6	3,9	—	—	—
3,9	4,2	—	—	—

Type 2 — Grown frames only

Depth, D, m (Motor)	Depth, D, m (Sailing and auxiliary)	Siding, mm	Moulding at heel, mm	Moulding at head, mm	Frame spacing, mm
1,5	1,8	24	31	24	205
1,8	2,1	34	40	31	230
2,1	2,4	42	50	37	255
2,4	2,7	52	61	46	280
2,7	3,0	62	74	55	305
3,0	3,3	72	87	65	330
3,3	3,6	81	100	80	355
3,6	3,9	90	117	98	380
3,9	4,2	100	140	117	405

Type 3 — Laminated frames only

Depth, D, m (Motor)	Depth, D, m (Sailing and auxiliary)	Siding, mm	Moulding, mm	Frame spacing, mm
1,5	1,8	25	25	205
1,8	2,1	31	34	230
2,1	2,4	37	43	255
2,4	2,7	43	51	280
2,7	3,0	50	61	305
3,0	3,3	57	74	330
3,3	3,6	62	87	355
3,6	3,9	69	105	380
3,9	4,2	78	126	405

Modulus of floors and frames, cm³

Depth, D, m (Motor)	Depth, D, m (Sailing and auxiliary)	Basic stiffener spacing, mm	$\frac{V}{\sqrt{L_{WL}}} \leq 3{,}6$ Floor at centre	Frame at side	$\frac{V}{\sqrt{L_{WL}}} = 5{,}4$ Floor at centre	Frame at side	$\frac{V}{\sqrt{L_{WL}}} = 7{,}2$ Floor at centre	Frame at side	$\frac{V}{\sqrt{L_{WL}}} = 9{,}0$ Floor at centre	Frame at side	$\frac{V}{\sqrt{L_{WL}}} = 10{,}8$ Floor at centre	Frame at side
1,25	1,75	380	35	15	45	15	55	20	60	20	70	25
1,75	2,25	395	80	30	105	35	125	40	150	45	165	50
2,25	3,00	410	150	55	195	60	235	70	275	85	305	95
2,65	3,35	425	245	85	320	95	395	115	460	130	515	150
3,00	3,75	440	375	140	485	140	595	170	695	195	—	—
3,50	4,25	455	560	210	725	210	890	250	1035	290	—	—
4,00	4,75	470	785	305	1020	305	1255	350	—	—	—	—
4,50	5,25	485	1060	425	1380	425	1695	470	—	—	—	—
5,00	5,75	500	1395	575	1815	575	2225	615	—	—	—	—

Fig. 23.5 The scantlings for GRP, steel and alloy, wood and composite yachts and small craft can be obtained from tables in Lloyd's Rules, two being illustrated above for wood (top) and GRP (bottom) transverse framing. They are only presented as examples and should not be used out of context with the Rules. (*Courtesy Lloyd's Register of Shipping.*)

24

SLOP CHEST

Ancillary Elements of Yacht Design

The slop chest was an old seafarer's box that contained his bric-a-brac and treasured belongings. Now our design is near complete we are going to delve into our design slop chest to see what are the creative efforts.

When engaged in the peripheral areas of marine design I always had to call upon the expertise of piping, mechanical and other engineers, as well as other sources such as previous plans, company data, manufacturers' brochures, or even consultation with a gnarled workman from the yard. Until experience is gained it is always a teeth-gritting task to try to design and draw a piping system or wiring diagram for a boat, or work out ventilation requirements and scheme the ducting. Fortunately, yachts and boats being very small in comparison to ships, the demands in these areas are not so exacting; highly theoretical calculations or intricate drawings are really not necessary. Nevertheless, whatever does emerge from the slop chest will require the reader to follow through in various ways, in similar fashion to my own explorations. The design topics unearthed from the chest can only be sketchily described, but will indicate some form of procedure to follow or route to pursue.

The first thing to emerge and be aired, being very ironic, is ventilation. In a previous chapter on equipment the various vents available were illustrated; now we want to delineate design requirements. The major purposes of ventilation are to:

(a) remove heat generated,
(b) supply oxygen for supporting burning, and
(c) remove odours.

The factors which affect human comfort are air temperature, humidity and purity.

Basically, what is required is a recommended input of about 50 cu.ft per man per minute for sleeping spaces, toilets and similar, which should be an ample allowance. Other spaces that are occupied intermittently will require about half the above figure.

The position of supply trunks should be pointed away from orifices to ensure good circulation of air because the air will travel a long way in a straight line if so directed. When supplying cold air, they should be directed horizontally, just beneath the beams; when supplying heated air, they should be directed downwards, and it may be necessary to have a shifting mouthpiece whose direction can be changed. Exhaust trunks should have their orifices as high as possible, while inlet and exhaust positions should be located at opposite ends to each other.

The efficiency of an installation is improved if bell mouths, to reduce the air velocity, are fitted. For the exhaust ducting these can have a short taper (about 1 in 2), while for the supply, a gentle taper (not more than 1 in 12) is required.

The calculation and drawing procedures require consideration of whether the ventilation is to be natural or forced (requiring fans), the losses due to friction and bends in trunking, etc. For our DIY purposes the basis design, guidance plans or brochures – as well as reference books, if necessary – should provide us with enough information to draw a ventilation arrangement.

Lighting for yachts and boats, as with most structures, is a combination of natural and artificial sources. Natural lighting is supplied by portholes and windows, while artificial lighting is from electrical power or possibly lanterns. Lighting systems for yachts and boats can be rather ambiguous from the design point of view, but here are some pointers and a suggested value for natural illumination, and some guidelines for other requirements. For cabins, toilets and similar accommodation spaces, 7 lumens/sq.ft (1 lm/sq.ft = 1 foot-candle) or 75 lumens/m^2 (1 lm/m^2 = 1 lux or 1 metre-candle) is the recommended intensity of illumination. You will need to refer to appropriate handbooks or reference books to interpret and apply these figures, but it may be said that, for simplicity, natural light should be sufficient in such spaces to read a newspaper, wherever situated. Artificial light is also a flexible area of design and will depend on the taste or desires of designer/owner. There needs to be a prime mover to generate electricity on board and this may come from an installed engine, or some other source such as batteries. A wiring diagram will need to be drawn on the plans of the boat, showing the layout of the system.

Another item from the slop chest of design covers the piping systems to be installed. It is not possible to cover all design features in this brief description, but one of the first steps is to plot the demands of fluid for the system by plotting an economical piping layout compatible with the layout and distribution of the yacht spaces to be supplied. Pipe sizes are allocated on a trial basis, allowing velocities of about 5–7 f/s. From this network is estimated the pressure required from the pump to create the flow. A simple diagram is shown in Fig. 24.1, and by working back from presumed simultaneous demands, the flow at different points may be found by simple addition, and pipe sizes allocated. The pump must be capable of driving the water or liquid from the outlet and delivering it to the remotest part of the yacht. Of course, the system design will involve calculation of flow, fittings such as valves, contractions and expansions, filters, nozzles, etc. Guidance from the basis design, or else brochures and other sources may be of great assistance. One last point: for yachts with some form of oil fuel system, this

Slop Chest

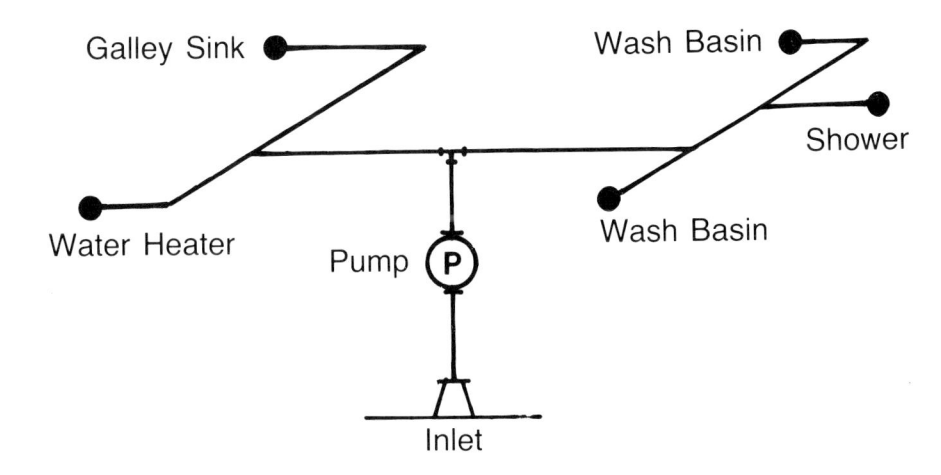

Fig. 24.1 Piping systems are part of our slop chest of yacht design. They require a layout and estimation of pressures involved, as illustrated above.

Table 24.1 Density of Commonly used Woods

Wood	Density lbs./ft^3	Wood	Density lbs./ft^3
Ash, English	44	Hazel	54
Ash, American	30	Lignum-Vitae	78
Beech	44	Mahogany, Cuba	48
Birch	47	Mahogany, Honduras	41
Cedar	31	Mahogany, Mexican	42
Elm, English	35	Oak, English	51
Elm, Canada	47	Oak, White (U.S.)	61
Fir, Spruce	30	Pine, Oregon	37
Greenhart	62	Pine, Pitch	41
Hawthorn	57	Teak	50
Elder	43	Yew	50

Table 24.2 Mechanical Properties of Glass

Material	Density lbs/in^3	Tensile Strength p.s.i. $\times 10^3$	Compressive Strength p.s.i. $\times$ in^3
Glass	0·08–0·15	20·30	300–500
Tempered Glass	0·08–0·15	40–50	300–500

needs to be decided upon at the early stages of design as it could affect stability and needs to be considered in this respect. Sewage and waste disposal systems also need to be looked into under piping systems.

For mild steel and aluminium alloy hulls (and wood or GRP hulls with metalwork below waterline) cathodic protection needs to be considered, as electrochemical corrosion occurs when two dissimilar metals are present in an electrolytic medium such as sea water. If another metal higher up the electrochemical scale is placed nearby, corrosion will occur on this metal and not the one to be protected. Sacrificial anodes such as pure zinc plates, magnesium or alloy, may be placed near the object to be protected, or else a very efficient form of cathodic protection is *impressed current protection*, used for large areas such as steel hulls, where a direct current is impressed in a circuit.

Finally, to conclude this chapter some formulae and tables are provided to determine such design aspects as chain locker size, material properties and other useful data.

Table 24.3 Mechanical Properties of Mild Steel land Aluminium Alloy

Metal	Density lbs./ft^3	Tensile Strength tons/in^2
Aluminium alloy	180	17
Mild steel	490	28

Conversion Factors

1 in = 2.54cm 1cm = 0.39 in.
1 foot = 30.5 cm   1 metre = 3.28ft.
1 yard = 0.914 metres   1 metre = 1.09 yards.
1 statute mile = 5,280 feet = 1.61km   1km = 0.62 statute miles
1 nautical mile = 6,080 feet = 1.85km   1km = 0.54 nautical miles
1 knot = 1.69 ft/sec. = 1.15 m.p.h. = 0.515 metres per sec.
1 pound (lb) = 0.45kg.   1kg. = 2.2 lb.
1 ton = 1.02 metric tonnes.   1 metric tonne = 0.98 British tons.
1 metric horsepower = 75 metres–kg/sec. = 0.99 British horsepower
1 British horsepower = 550 ft.–lb/sec = 1.01 metric horsepower

Table 24.4 Stowage of Chain Cable (Cubic Feet per 100 Fathoms)

Dia. (ins)	5/8	3/4	7/8	1	1 1/8	1	1 1/4	1 3/8	1 1/2	1 5/8	1 3/4	1 7/8	2
Cubic ft.	14	20	27	35	44		55	66	80	93	105	130	160

25

TIED UP AT THE PORT OF DESIGN
Final Calculations

We have reached the end of the spiral – it now remains to tie up at the port of design.

The calculations can be gathered and presented to form a comprehensive description of the yacht in various conditions. This very extensive exercise is obviously not essential to our self-design effort, but it does add to our knowledge of the design calculations and is a useful adjunct to our total knowledge of yacht design.

The first – and probably most important – set of calculations is presented on what is called the *Displacement Sheet*. This is but an extension of our previous simplified calculations for displacement, centre of buoyancy, centre of flotation, etc., and is set up in a compact, standard tabular form. Using the waterlines offers a check on the clerical accuracy of the calculation.

My own efforts during the years in calculating a Displacement Sheet have been rather chequered. There were times when I have 'smiled' through the figure work and emerged quite untroubled. Every column has added up nicely and cross-checking has been right on the nail. Then there have been torturous occasions when, with gritted teeth and furrowed brow, I have had to plough repeatedly through the endless and merciless numbers which refused to comply with each other or provide a union of rows and columns. Chasing a spare hundred or two or trying to track down an anomaly in the cb figure can be a thankless task.

A very important calculation chart is called the *Hydrostatic Curves* (Fig. 25.1). It is exactly what its title suggests, being a set of curves that deal with hydrostatic elements of the yacht, and is drawn to a vertical scale of draught. Hydrostatic properties such as displacement, centre of buoyancy, metacentric height, etc., can be read off at any waterline and thus provide a complete picture of the yacht for any condition at which it may float. To draw this set of curves one has to calculate and plot the various hydrostatic values at each waterline and then draw a smooth curve through the points.

I once drew a set of Hydrostatic Curves for quite a large vessel and glowed with pleasure when viewing the smoothness and accuracy of the completed work. The pleasure was short-lived; it was pointed out to me that I had drawn the draught or waterline scale for the wrong bottom of keel. Ah well! An eraser and an hour or two of unpaid overtime eventually put the chart right.

Design Your Own Yacht

A very useful set of curves that aid the other calculations are the *Bonjean Curves* (Fig. 25.2). They are plotted by taking the sectional area at each station – taken up to each waterline level – and plotting these horizontally (the *abscissa*, as such horizontal plots are called) from the centreline at each relevant waterline.

This set of curves is a useful aid for the other calculations and curves.

It is unlikely that our DIY design will require the following calculations but they are worth mentioning because they could very well apply to large,

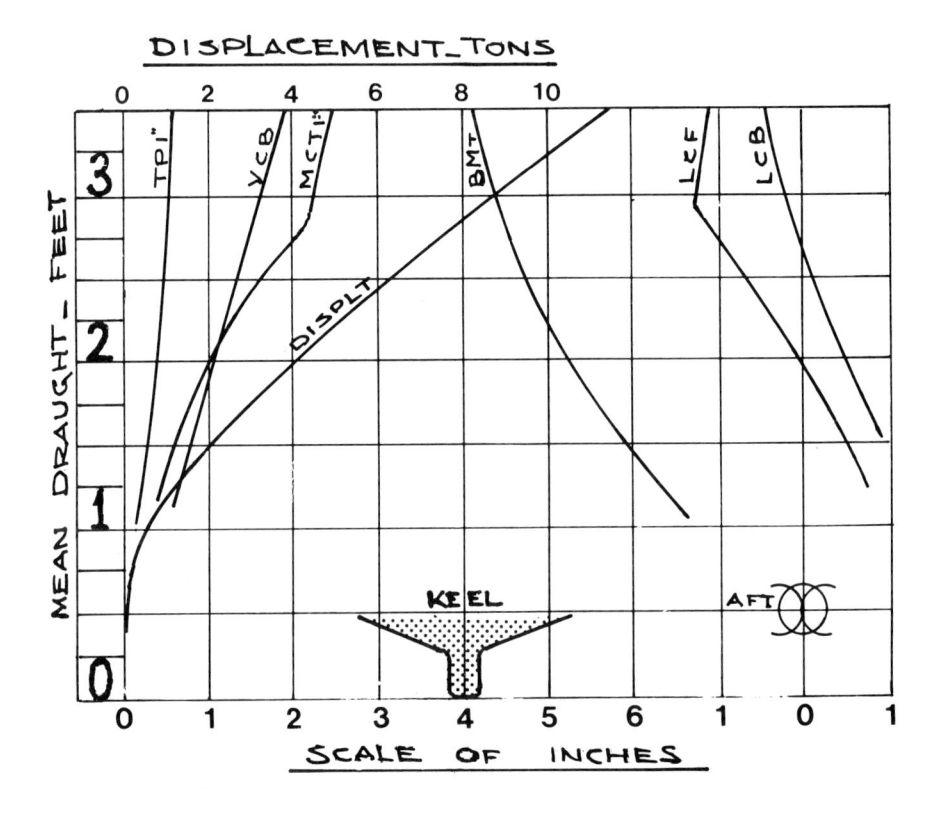

TP1" – Tons per Inch Immersion (1" = one ton)

V.C.B. Above Base (1" = one foot)

M.C.T.1" – Moment to Change Trim One Inch (1" = 0.4 ft–ton)

BM_T – Metacentric Height (1" = one foot)

L.C.F. – Centre of Flotation about Amidships (1" = two feet)

L.C.B. – Centre of Buoyancy about Amidships (1" = two feet)

Fig. 25.1 An important graph that ties up our hydrostatic calculations is called the Hydrostatic Curves, as illustrated above.

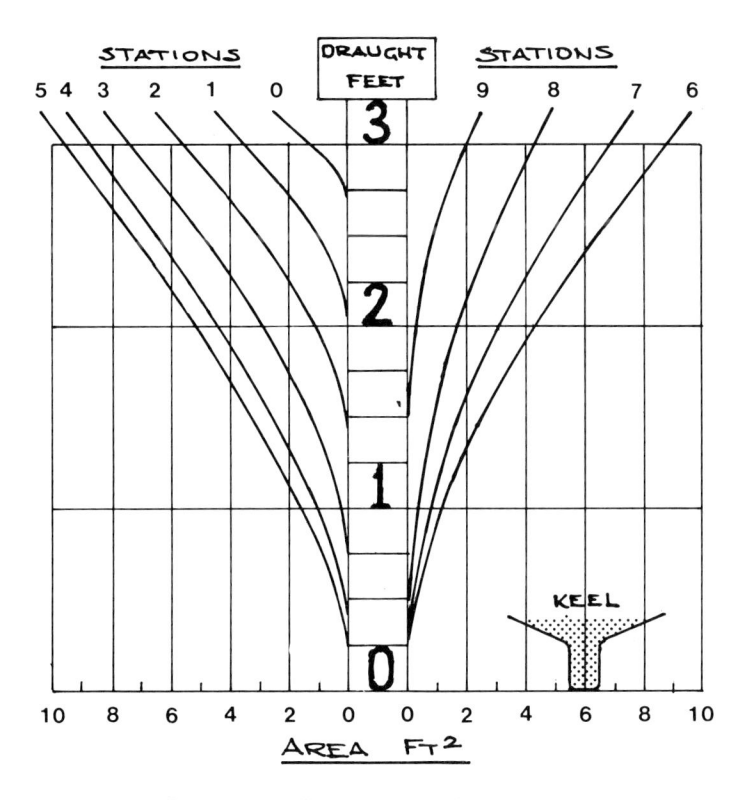

Fig. 25.2 Bonjean Curves assist other yacht calculations, giving the areas at each station up to each waterline. A set of Bonjean Curves is shown above.

ocean-going yachts, which may be the design fantasy of some readers. For vessels of a certain size there is usually a *minimum summer freeboard* based on official regulations and rules. This standard freeboard is corrected or adjusted according to vessel type, service and certain design factors. There is also a *standard sheer* and *round of beam* which may also be corrected for similar reasons.

The subdivision of hull also has to be looked into for large yachts and boats, and a standard *Floodable Length Curve* may have to be prepared. Again, rules govern the calculation, together with corrections and various conditions of flooding, and flooded stability may have to be investigated in the process.

Most large yachts and craft may require some form of launching calculations if they are built on a slipway. A drawing of the hull profile is placed on a drawing of the slipway and, using the Bonjean Curves, it is moved down the slipway in order to calculate the various buoyancies as it enters the water. Moments and pressure on ways are also calculated, along with other factors such as tipping moment, pressure on fore poppet (the structure built under the hull at which the hull will hinge when it takes to

Plate 28 Many yacht calculations can now be performed on computer and reputable marine bodies provide programs and services covering all elements of such calculations.

the water), release arrangements and sliding friction during the launch. A set of *Launching Calculations* and *curves* is carried out in respect to this aspect of design.

Nowadays, yacht and boat calculations can be performed on a computer and one of the many programs offered is a hydrostatics and stability package by the Wolfson Unit at the University of Southampton, Hampshire, England. This Unit has gained an international reputation for its consultancy services to the marine industry and can supply programs that enable naval architects and yacht designers to perform accurate calculations over a range of design problems. Lloyd's Register of Shipping also offers a service for computerised design calculations.

One final aspect in wrapping up our self-design is whether the yacht is to be built under racing rules. The Royal Yachting Association (RYA) is one of the oldest and most prominent organisations in the world concerned with the regulation and classification of racing yachts, and two very popular types of sailing yacht have been selected to illustrate how class rules apply: these are the International Flying Fifteen and the Albacore dinghy. For the International Flying Fifteen a measurement certificate is obtained (if the builder pays the International Class fee to the National Flying Fifteen Association) to prove that the boat is measured by an approved class measurer, who will then complete the measurement form, which is then sent to the owner's certification authority, together with the International Class Fee receipt and any registration fee required. The measurements are very comprehensive and detailed, covering the hull, transom, keel, rudder, and weight of yacht. It also covers the various aspects of sails and rigging, and over one hundred are listed in the measurement form, whose seven pages have three columns requiring minimum, actual and maximum measurements for a particular rule number. The Albacore – which is a restricted class – has similar requirements. The RYA provides international rules for this dinghy, with national variations.

It is not possible to cover the multitude of rules and requirements for class racing of yachts when considering a design, so if you intend to design such a yacht it is essential that you equip yourself with the relevant rules with which you should comply.

So, having tied up at the design port, your next venture into yacht design could very well enter into deeper water – the subject of the next chapter.

26

INTO DEEPER DESIGN WATERS
More Sophisticated Elements of Yacht Design

Having covered the essentials of yacht design, the reader may wish to enter deeper design waters, if only to be more informed on the subject.

In the past I often felt that the journey in marine design was an endless one; one would get caught up in the infinite web of alternatives that a change in dimensions can produce, tangling with the stability implications, resistance considerations and space requirements. For example, alter the beam of a yacht and the original web of design becomes more enmeshed. Not only that, there is continuing research that informs us of new or changed views when shuffling the design parameters around, quite apart from the rapid advances being made in marine equipment, materials and their protection. It is almost certain that revolutionary concepts in hull design will emerge in the future to sail these deeper waters, while theory will become an increasing drive toward yacht creation. No wonder this chapter can only skim the surface of deeper design waters in the hope of advancing the reader's knowledge.

First, a dip into the more theoretical aspects of the calculations, beginning with resistance and powering. When Froude advanced his theories on model tank experimentation he was able to show that the wavemaking resistance and drag could be separated, and consequently conducted experiments with long planks to obtain the frictional component at his tank in Torquay, assisted by his son. The results, later modified, were put into non-dimensional form as what are known as circular coefficients of the various parameters. One such important coefficient for tank experimentation is given by:

$$Ⓒ_m = \text{Resistance (lb)} \times \frac{1.311}{V^3 \sqrt[3]{\Delta^2}}$$

where Δ = model displacement and C_m = total model resistance.
The British system is based on this, while US experimenters use the notation devised by a well-known American marine researcher, Admiral Taylor. His notation is of dimensional form where the resistance is expressed in lbf per tonf of displacement as a resistance coefficient R/Δ, and the displacement:length ratio in the form:

$$\text{Displacement:length ratio} = \frac{\Delta}{(L/100)^3} \; (\Delta \text{ in tonf})$$

Into deeper Design Waters

Plate 29 Tank experiments enter into the deeper waters of yacht design, especially in resistance, powering and propeller studies. Above is a view of British Maritime Technology's manoeuvring facility. (*Courtesy British Maritime Technology.*)

Care has to be taken when comparing the American notation to other results because of its dimensional context.

Regarding screw propulsion, I previously suggested a simple approach to obtaining a suitable propeller. There are a number of theoretically-based methods of analysing the thrust, torque, etc., but there are also results from tank experiments using the propeller both behind the hull and in open water. The thrust and torque obtained at an estimated speed of advance are then put into non-dimensional coefficient form. Propeller designers use these results in combination with curves obtained from many open water experiments on various propellers, called Methodical Model Series. Long and complicated calculations are involved thereafter, which it is not possible to enter into here.

Moving on to stability, there are a number of methods available in calculating the full range of stability, discounting computer programs. These include the Barnes Method and use of the integrator, but in any case

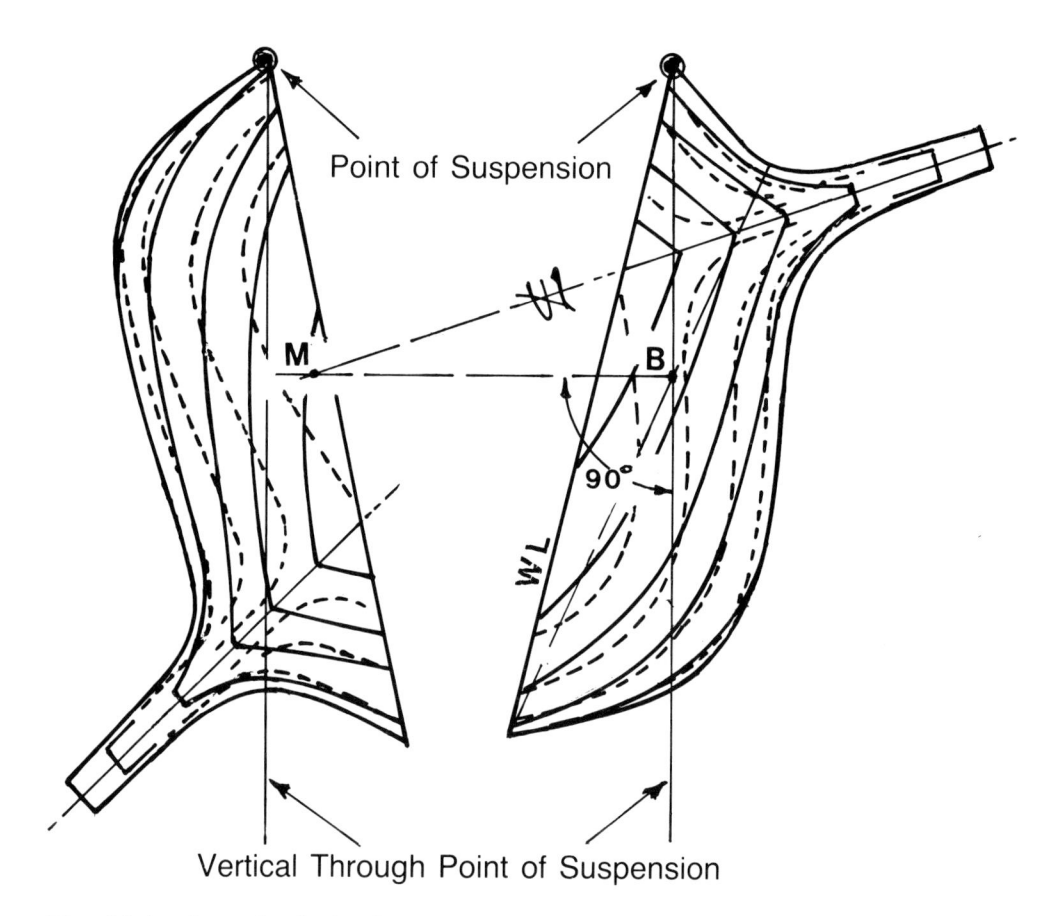

Fig. 26.1 A method of calculating the full range of yacht stability – known as Blom's Method – uses a contracted cardboard model of the hull (see above) to determine stability values.

elaborate procedures and calculations are involved. An experimental method known as Blom's Method is probably most suitable to this book. Briefly, it consists of making a contracted paper model of the hull by cutting out cardboard sections of the hull at each station and for each angle of inclination being considered, gluing them together in their correct positions, then paring them down parallel until they weigh the same as the upright model. The model is then suspended at each corner and a plumb bob is used to mark vertical lines in these suspended states, their intersection giving the cb (B) at the inclined waterline being considered. The line BM is then drawn at right angles to the inclined waterline to cut the centreline of the hull, and GZ is then scaled. The whole procedure is graphically illustrated in fig. 26.1.

A full range of yacht stability using the method just described will produce a GZ curve as shown in Fig. 26.2. A set of curves not previously mentioned is the Cross Curve of Stability. For this set of curves the stability

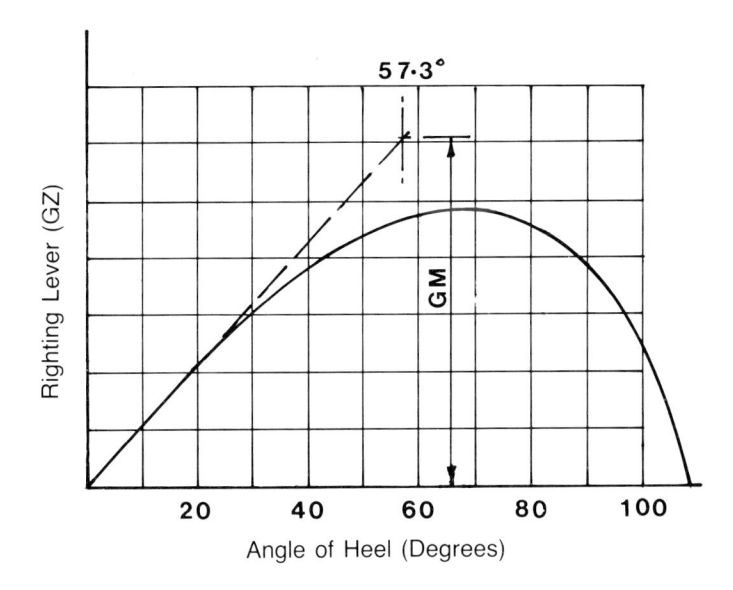

Fig. 26.2 The GZ curve above shows the type of curve produced when calculating the full range of stability for a yacht.

is drawn for different displacements over the range and a cross curve is constructed for each inclination, with GZ as the vertical coordinate and displacement as the horizontal coordinate or abscissa. The correct value of GZ for any required displacement can then be read off and used as an ordinate for a curve of statical stability at constant displacement.

Considering the linkage of form and transverse stability, the following considerations are generalisations that may help in your self-design. If length is increased in proportion to displacement – beam and draught being kept constant – the BM is also increased. If length is increased at the expense of beam there is a reduction of stability over the full range; if increased at the expense of draught, initial stability is increased but reduced at larger angles. Beam has the greatest influence on transverse stability and BM is increased when beam is increased at the expense of draught. Initial stability is also increased when there is a reduction in draught in proportion to a reduction in displacement; nevertheless, such stability is reduced for large angles. A change of displacement – length, beam and draught being kept constant – will generally increase BM.

Discussing hull form, a modern application (though old in concept) is the multi-hull. The shape of the monohull is what provides its stability, as against the multi-hull which relies on greater stability by splitting and spacing the hull form into two or three separate hulls, generally not requiring ballast. The *catamaran* has twin identical hulls, smaller types being sailed like dinghies with beams to tie the hulls together; larger versions have a continuous, rigid bridge deck between hulls, usually covered for accommodation purposes. The *proa* has a main hull and outrigger on the

Design Your Own Yacht

Plate 30 High speed craft such as Cougar Marine's 33'-0" US. 1-33 above, are specialist boats usually of vee-shaped hull form.

windward side for stability purposes and therefore cannot tack in the normal way. The *trimaran* has a central hull and two smaller wing hulls. The central hull is the working space, being built with living quarters on larger versions, and the leeward wing provides stability while the windward one rides clear of the water. The three types are illustrated in Fig. 26.3.

High-speed craft are specialist boats and their resistance is readily calculable. Their hull form is usually vee-shaped and has a hard chine. Such boats should run at the flattest angle that will afford the necessary lift in

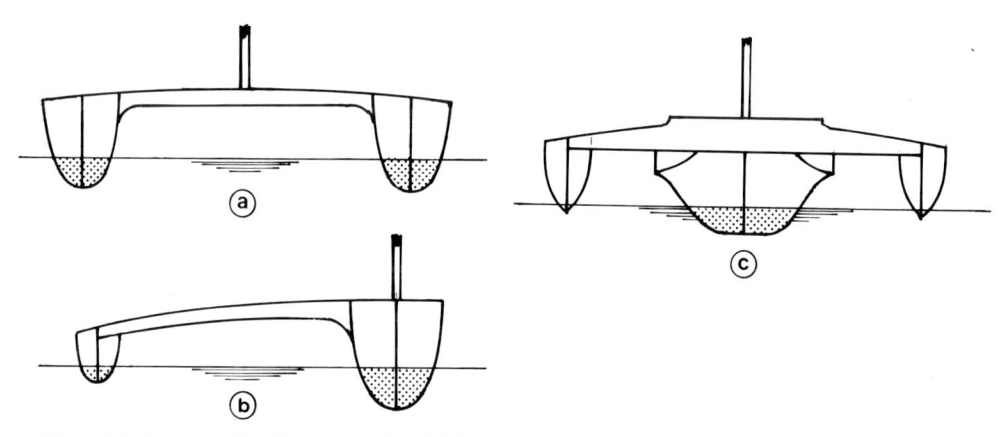

Fig. 26.3 Multihulls, though old in concept, are of recent innovation as sailing and pleasure craft. Figure (a) is a catamaran, (b) a proa and (c) a trimaran.

194

order to reduce residual resistance, yet at the same time obtain the steepest angle to leave sufficient surface for the lift. One version that avoids some of the problems created by the above contradictory requirements is the stepped hull, which nevertheless suffers other disadvantages as a seagoing craft.

The theoretical aspects of sail area and propulsion involve many problems not easily solved, or able to be presented in this book. In the Driving Winds chapter I presented a simple approach to obtaining the relevant design elements of sails. The present discussion, though involving more sophisticated theory, can only reach a certain point. The driving and drift forces on a sail are obtained by a combination of calculation and the use of standard curves. Let us take one uncomplicated example, leaving out the actual detailed arithmetic in the calculation. An efficient yacht should be able to sail within about four points (45°) of the true wind. Assume a yacht is sailing on this course at 5 knots. Its sail has an aspect ratio of 2.5 and an area of 100 sq.ft. Suppose the true wind acting is 15 knots, then by drawing or calculation the relative wind can be shown to be about 18.8 knots. Again, by calculation or measurement, the angle off the bow will be 34° and, assuming the sail is set midway between the course of the yacht and the relative wind, the angle of incidence and trimming angle will each be 17°, as illustrated in Fig. 26.4. The wind components acting normal to and along the sail can be found to be 5.5 knots (normal) and 18 knots (along) the sail.

Now data is available for the normal force coefficients on thin, rectangular plates, as shown in a simplified version in Fig. 26.5. Referring to this and using the formula for pressure acting as given by:

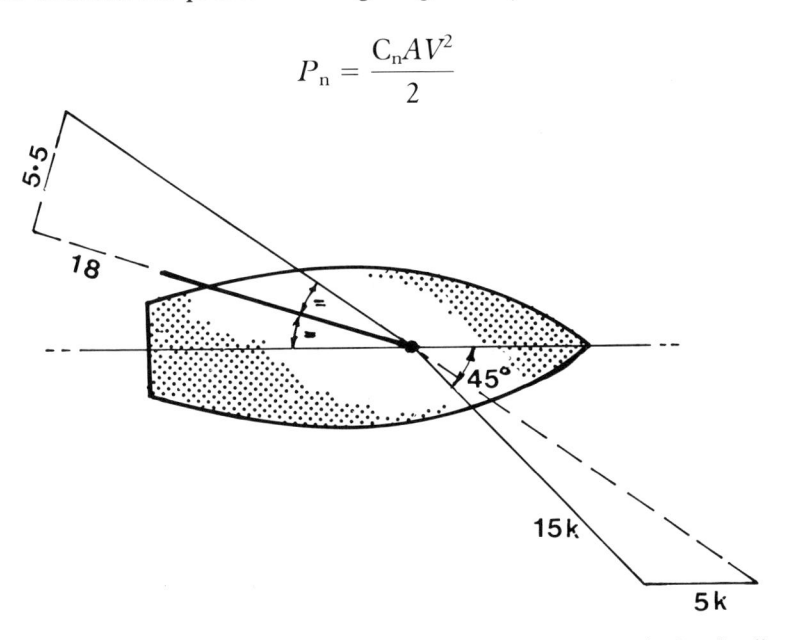

$$P_n = \frac{C_n A V^2}{2}$$

Fig. 26.4 A graphical or calculated approach can be made for finding the forces acting on a sail, as illustrated above.

Design Your Own Yacht

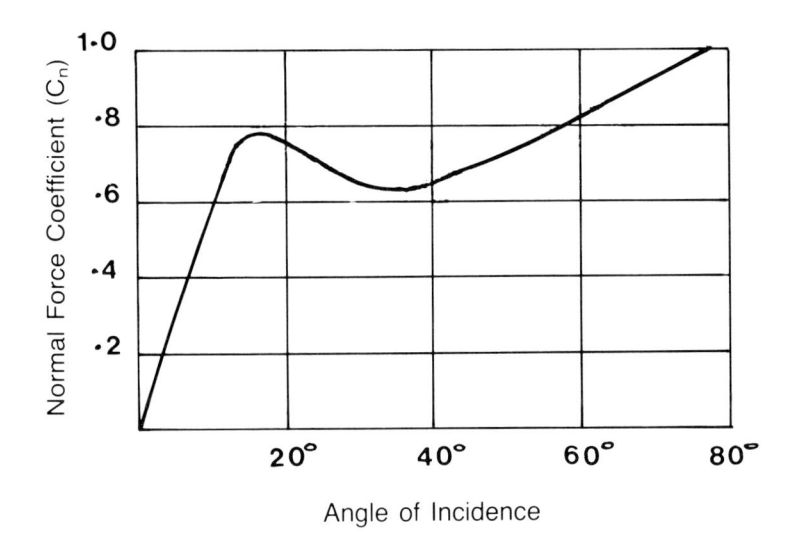

Fig. 26.5 Data and graphs are available to find coefficients that allow the normal forces on sails to be calculated. One such graph is shown above for an aspect ratio of 2.5.

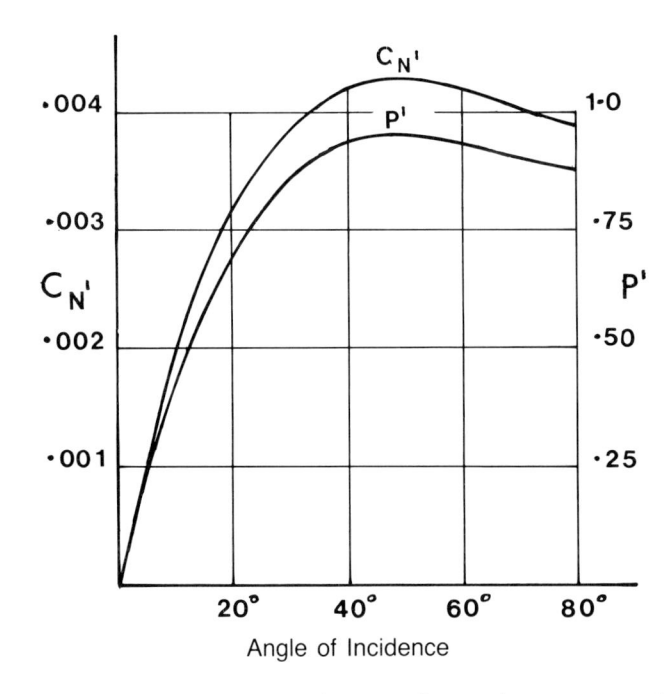

Fig. 26.6 The graph above provides an alternative means of calculating sail forces $C_N{}^1$ is applied in the formula $P_N = C_n{}^1 A V_k{}^2$ (V_k in knots). P^1 is wind pressure for a 15 knot wind at different angles of incidence and aspect ratio 2.5.

where P_n = force normal to foil, A = foil area, and V = wind speed in ft/sec.

The resulting unit pressure can be calculated to be 97 lb, using a value of about 0.8. Alternatively, another approach is to use the results illustrated in simplified form in Fig. 26.6, which will give a similar result. The formula to use is:

$$P^1 = C_n^1 A V_k^1$$

Where C_n^1 is an approximate normal force coefficient and P^1 wind pressure, both for a 15-knot wind, aspect ratio about 2.5. Using such values, the driving force will be found to be 28 lb and the drift component 92 lb. The course taken by the yacht under such conditions will be the ratio of these components where V_k is in knots.

The problem now becomes more difficult, considering the many variables. For instance, sails are not rigid plates but bend in face of the wind; rigs which split the sail plan between two masts lose in efficiency. Sail area may also be governed by rules and regulations concerning Class; in any case it may be said that to achieve highest efficiency a sail plan has to achieve the best lift:drag ratio practicable under conditions ranging from a fine angle of attack to the fully stalled state. Reassuringly, an effective driving force will be obtained from a modest sail area. Further analysis of this topic must terminate at this point and reference should be made to specialist information and data – not possible in this book – in order to make deeper investigations on design.

The contours of ocean waves play an important part in marine design, including those aspects concerned with resistance and strength. One wave

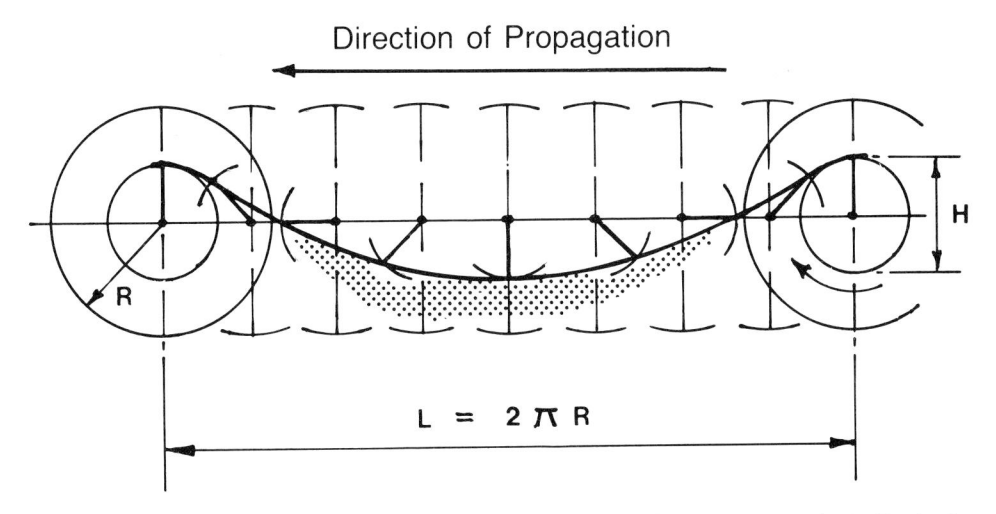

Fig. 26.7 The trochoidal wave form above closely corresponds to that of sea waves and is used in the strength calculations to balance yacht hull.

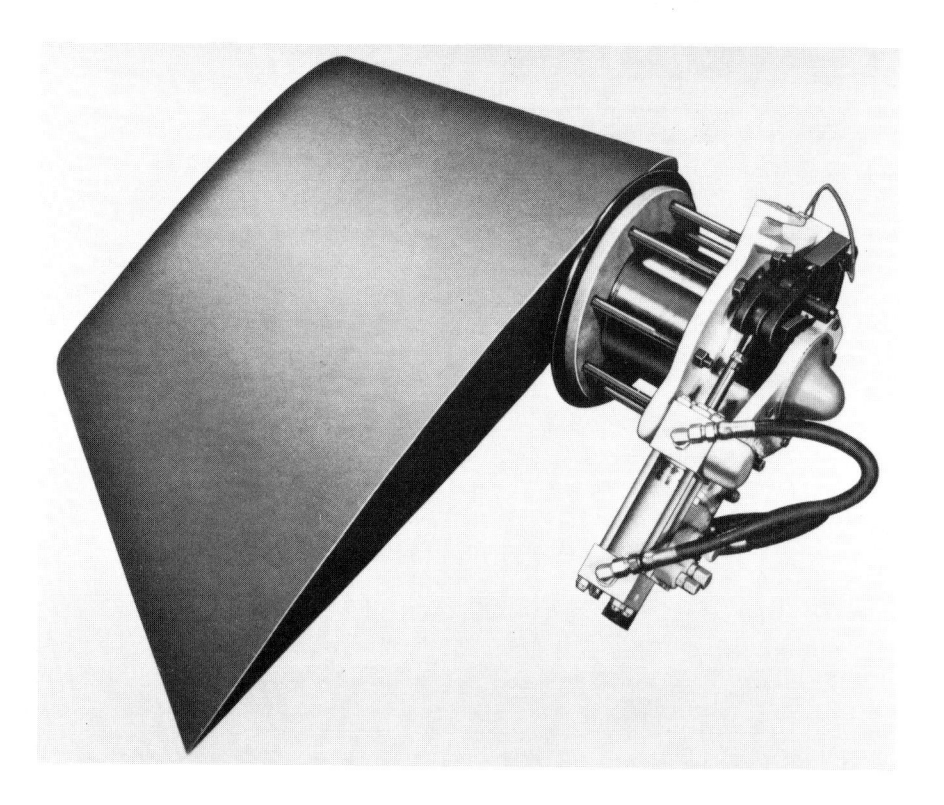

Plate 31 Sophisticated equipment and instruments can now be bought when entering the deeper waters of design. For instance, roll stabiliser systems as that shown above assist in damping the rolling motions of a yacht. (*Courtesy NAIAD.*)

contour used for the strength calculation is called the trochoidal wave. Imagine a point on a wheel rolling on a horizontal surface. The path the point will trace as the wheel rolls is of trochoidal form, as illustrated in Fig. 26.7. Tables (found in a suitable handbook) give the coefficients for construction of a trochoidal wave used in the strength calculation and typically, for such a calculation the height:length ratio is 1/20.

One area not yet touched upon in this book is vibration. Assessing the vibration for both small and large vessels is not an easy task at all and while there are empirical formulae for large ships, very little data can be obtained for small craft such as yachts and motor cruisers. Like all similar structures a hull is subject to natural frequencies which may produce undesirable results if they synchronise with an exciting force, say, such as engine revolutions. This produces resonance and to avoid this, communication of the exciting force must somehow be prevented. Vibrations may occur vertically, horizontally, torsionally or longitudinally. The natural frequencies of vertical vibration depend chiefly on depth and the rigidity of fore and aft constructional members; those in the horizontal plane, on the beam and rigidity of transverse construction. Unfortunately, the effects of vibration can only be discovered after a yacht is built – unless long and complex calculations are carried out – but a cure can be found by strengthening those areas affected.

To conclude this chapter, let us now look at some sophisticated aspects of equipment and outfit. This field of yacht design is rapidly advancing, especially in the areas of navigation. For instance, a total package can be obtained for the luxury end of the market, but no doubt such a system will be well within the average yacht owner's pocket in the future. To cure adverse rolling, stabilisers can be fitted on yachts, while for hull protection there are such royal coatings as epoxy resins, which require more care in application but offer excellent protection. There are many sophisticated paints on the market offering less resistance of hull etc., while one has to also consider cathodic protection.

As a parting word – *bon voyage* in your voyage of yacht self-design.

INDEX

Page numbers in bold type indicate an illustration

A

above-deck profile, 67, **67**
accommodation space, 50, 63, **64**, 119, 122–3, **133**
Admiralty formula, 107
advanced design techniques, 190–99
aerodynamic force, coefficient of, 101
aluminium alloy, 31, 36, 54, 132, **180**, 184
amidships, 8, **9**
analysis pitch, 106
angle of entrance, 11, 96, **96**
angle of incidence, 98–9
angle of vanishing stability, 158
apparent slip, 104, **105**
area calculation, 69–72, **70**

B

balance, 101
basic multipliers, 69
basis design, 93, 94
battens, 18, 19, **20**, 137, **139**
beam, 8, **9**, **12**, 13, 14, **15**, 91, 93, 94, 95
Bending Moment Diagram, **167**
bending stress, **165**, 167
bent timber construction, 32, 33
Bermudan sails and rigging, 37, **37**, **38**, 39, 40
Bernoulli, Daniel, 98, 103
bilge diagonals and keels, 11
Block Coefficient, 14, **15**, 16, 90, 91, 94
blocks, 47, 48, **49**
Blom's Method, **192**
Body Plan, 7, 9, 21, 135, 136, **139**, 145–6
body resistance, 84, 85
bold bow, 63, **65**
Bonjean Curves, 186, **187**
boss (of propeller), **105**
boundary layer, 81
bow lines, 10, **10**, 136, **137**, 138
bow shape, 63–4, **65**
bow wave system, 83, 84, **84**
box scale, 22, 23
brackets, **12**, 13
brake horsepower (bhp), 107
broken sheerline, 63, **64**

Bulkhead Plan, 172, 173
bulkheads, **12**, 13, **13**, 34, 117–18, 171
buoyancy, 147, 152, 166, *see also* displacement
Buoyancy Curve, 166, **167**
buttock lines, 7, 10, **10**, 136, **137**, 138

C

camber, 8, **9**, 139, **140**
canoe stern, 64, **65**
carvel construction, 13, 32, **32**, **34**
cat-boat, 39
cavitation, 111
centre of area, 71, 72, 75, **75**, **102**
centre of buoyancy (cb), 59, 72, 74, 147, 151, 154
centre of effort (ce), 101, **102**
centre of flotation (cf), 161
centre of gravity (cg), 59, 74–5, **74**, **75**, 76, 147–9, **148**, 151, **152**, 154, 159
centre of lateral resistance (CLR), 101
centreboard, 26, **26**, 27, **66**
Classification society rules, 32, 58, 59, 61, 117, 133, 149, 170, 171, 174–80
cleats, 48, **49**
clew, **37**
clinker construction, 13, 32, **32**, **34**
clipper bow, 63, **65**, 88
clutches, 44, **45**
coefficients (hull design), 14–16, **15**, 88, 90, 101
common interval (CI), **70**, 72
compasses (for drawing), **20**, 22
compasses, navigational, **46**, 48
components of forces, 99, **99**, **128**
compressive stresses, 165, **165**
Construction Plan, 172, **173**, 178
constructional plans, 171–80
contours of design, 62–7
counter stern, 64, **65**
couple, 152
critical speed, 81
cruiser stern, 64, **65**
cruisers, 26–7, **27**, 34, 67
curve of statical stability, 158
curves, 18, 19, **20**, **21**, 69–72, 137, 138, *see also* graphical data

200

Index

cutter rig, **38**, 39
cut-up, 11

D

deadwood, 11
deck beams, 13
deck equipment, 47, 48
deck girders, **12**
Deck Plan, 172, 173, **178**
deep displacement, 52, **52**
density, 73, 143, 183
depth, 8, **9**
design instruments, 17–24
design procedures, 56–61, 57
developed horsepower (dhp), 107
developed outline of propeller, **105**, 106
diesel engines, 41, **42**
dinghies, 26, **26**, 34, 67
disc area ratio (DAR), 105
displacement, 14, 59, 93, 94, 141–7, 151, 166, 186
Displacement Sheet, 185
displacement: length ratio, 190–91
displacement-type hull, 52, **52**, 66, 83, 84, 107–9
divergent waves, 82, **82**
double chine, 52, **52**
drag, 81, 82, **82**, 84–7, **84**
drag force, 99
draught (draft), 8, **9**, 11, 14, **15**, 16, **91**, 117
drawing instruments, 17, 18, **20**, 21–4
drift, 99, 127
drive arrangements, 4, **45**
driving component of wind, 99
dryness of deck, 63, **64**

E

eddies, 82, 87
effective horsepower (ehp), 107
effective pitch, 106
Elevation, 10, 116, 118, 121, 135, 136, **139**, 173
elm timber, 30, 132, 183
engine bearers and seatings, 11, **12**
Engine Seating Plan, 172
engines, **12**, 41–5
entrance, 11
equipment, 46–50, 55
explanation of terms, 7–16

F

face pitch, 105–6
fin, 66, **66**
fittings, 46–50, **49**, 55
flare, 10, **11**
flat scale, 22, 23
flats, 117, 119, **122**
Floodable Length Curve, 187
floors, **11**, **12**, 13, 117, 171
flow characteristics, 80–87
foot (of sail), **37**

forces, 73, **74**, 99, **99**, 167
foresail and forestaysail, **38**, 39
form characteristics of design, 80, 88–97
form coefficients, 14–16, **15**, 88, 90, 91
four-stroke engines, 41
frame brackets, 13
frames, **12**, 13, **13**, 171
free surface effect, 159
freeboard, 9, **9**, 63, **64**, 187
frictional resistance, 81, **81**, 84, 86–7
Froude, William, 85–7, **86**, 92
fullness of shape/waterplane, 90

G

gaff, 36
gaff rig, **38**, 39
gaff sail, 37, **37**
gasoline engines, 41–2
gears, 44
General Arrangement Plan (GA), 23, 59, 93, 94, 114–26, **125**
Genoa foresail, **38**, 39
geometric pitch, 105–6
geometry of hull design, 9–11, 58, 95–7, **95**, **96**
glass, 183
glass reinforced plastic (GRP), 13, 30, **31**, 43, 54, **180**, 184
graphical data, 70–72, 84–5, 90–92, 100, 108–10, 112–13, 145, 148, 167, 185–8, 192, 195–7

H

half-sections, 9, 137, 138, 143
halyards, 35, **36**, 37
hard-chine hull, 52,52
head (of sail), **37**
headsail, 37
heavy displacement, **52**, 63, 66
heeling lever, **102**
history of sailing, 2–6, 35, 47, 85, **115**, 116
hogging condition, 164, **164**, 165
hollows and humps, 83
horsepower, 106–13, **106**
hull design
   calculations, 14–16, **15**, 59, 60, 68–79, 88–97, 134–80, 185–9
   flow and resistance, 80–87
   form characteristics, 80, 88–97
   geometry, 9–11, **10**, **11**, 58, 95–7, **95**
   structure, 11–13, **12**, 43, **60**, 61, 165–6, **178**
   types, 52–3, **53**, **58**, 62–7, 193–5, **194**
hump speed, 83, 84
hydrostatic curves, 60, 185, **186**
hydrostatic data, 59

I

immersed volume, 14, 16
inboard engines, 42, **42**, 118
inboard/outboard engines, 42, 43, **43**

201

Index

inertia, 71, 76, 156, *see also* moments of inertia
instruments for drawing the design, 17–24
introduction, 1–6
iron construction, 32

J

jargon, 7–16
jibs, **38**, 39

K

keel, 11, **11**, **12**, **13**, 27, 66–7, **66**, 177
ketch rig, **38**, 39, 40
kinetic energy, 98
knuckle, 11

L

laminar film, 81, **81**
laminar flow, 80, 81, **81**
laminated construction, 34
Launching Calculations, 187, 188
Laws of Motion, 73
lead, 101, **102**
lee helm/rudder, 101
leech, **37**
leeway component of wind, 99
left-hand propeller, 104
length, 8, 57
   between perpendiculars (LPB), 8, 9, **9**, 13, 117
   waterline *see* waterline length
level lines, 9, 136
lift force, 99
light displacement, 52, **52**, 63, **64**, 66
lighting system, 49, 182
Lines Plan, 10, **10**, 23, 24, 59, 80, 95, 96, 106, 116, 117, 134–40, **137**
Lloyd's Register of Shipping, 32, 117, 133, 149, 170, 172, 174–80, 188
LOA *see* overall length
Load Diagram, 166, **167**
lock-train gears, 44
longitudinal stability, 155, **156**
longitudinals, 11, **13**, 171
loose foot (of sail), **37**
luff, **37**
lugger (lug-sail), **37**, **38**, 39
lugs, 13
LWL *see* waterline length

M

mainsail, 37, **38**, 39
maintenance tools, 50
manufacturers' brochures, **54**, 181
masts, 35, 36, 48, 119
materials, 30–34, 54, 164–5
maximum moulded breadth, 8, **9**
measurements (hull design), 14–16, **15**
metacentric height, 59, 155, 156, 157
Midship Coefficient, 14, **15**, 94

Midship Section, 172, 173, **174**, **178**
midship-section area, 14, **15**, 94, **168**
mild steel, 32, 54, 132, 184
mizzen, **38**, 39
moments, 59, 71, 73–5, **74**, 101, **102**
   of inertia, **69**, 71, 76–7, **76**, 156–9, 168–70
Morrish Formula, 91, 93
motion and stability, 160–63
motor craft, **28**, 63, 67, 103–13
motors *see* engines
moulded dimensions, 8, **9**, 14, **15**
multihull, 193–5, **194**

N

navigation equipment, **47**, 48, **48**
negative/neutral stability, 154, **155**

O

oak timber, 30, 33, **33**, 183
ocean-going craft, 27, **28**
offsets, 23
ordinates, 69–72
outboard engines, 42, 43, **44**
overall length, 8, **9**, 95
overhanging bow, 63, **65**
overhangs, 63–4, **65**

P

paper for drawing, 23–4
pear shapes, 19, 21, **139**
pencils, 23
petrol engines, 41–2
physical laws, 73
pillar, **12**
piping system, 50, 181, 182, **183**
pitch (of propeller), 104, 105–6, 109
pitch pine timber, 33, 132, 183
pitching, 155, **156**, 160
Plan View, 10, 21, 116, 118, 121, 135, 136, **137**, **139**, 173
planimeter, **22**, **142**
planing-type hull, 52, 83, 109
planning a procedure, 56–61, **57**
plans, 23, 59, 114–26, 134–40, 171–80
plastics, 30–31
plumb bow, 64, **65**
plywoods, 34
polyester resins, 30–31
positive stability, 154, **155**
potential energy, 98
power in relation to speed, 107–9
powering, 53, **53**, 58, **58**, 80
pressure, 73, 98, 130–31
pressure difference, **98**, 127
Prismatic Coefficient, 14, **15**, 16, 90, **90**, 161
procedures for design, 56–61, **57**
profile of yacht, 62–7
projected outline of propeller, **105**, 106

202

Index

propeller design, 103–13, **105**, **110**, 191
purpose of yacht, 52

R

race (of propeller), 104, 129
racing associations and regulations, 188–9
racing craft, 26, 27, **28**, 67
radar and radio equipment, 48, **48**
Radius of Gyration, 76, **76**
rake of propeller, **105**, 106
real slip, 104, **105**
reduction gears, 44, **110**, 112–13
relative wind, 100
repair tools, 50
reserve of buoyancy, 63, **64**
residuary resistance, 87
resistance and flow, 80–87, *see also* drag
reverse sheer, 63, **64**
rigging, 35–40, **36**, **38**, 48, **49**
right-hand propeller, 104, **105**
righting lever, 155, 156
rise of floor, **11**
roach, **37**
ropes, 47, 48
round of bilge, 7, **11**
round up, 8
rudder design, 54, 127–33, **128**
runner, **36**
running rigging, 35

S

safety equipment, 49
sagging condition, 164, **164**, 165
sail area, calculation of, **100**, 101
sail design, 35–40, **37**, **38**, 48, **58**, 98–102, 195–7
sawn timber construction, **32**, 33, **33**
scale rulers, **20**, 22–3
schooner rig, **38**, 39, 40
screw propeller, **110**
seating girder/top plan, **12**
Second Moment of Area, 76
sections through hull, 9, **10**, **12**, 116, 118, 121, 173, **174**
semi-displacement type hull, 107–9
separation of flow, **81**, 82, **82**
set squares, 18, **19**
shaft horsepower (shp), 107, 108–9
shaft line, **12**
shearing forces, 166, **167**, 168
sheer, 8, 63, **64**, 187
sheets, 35, **37**
shrouds, 35, **36**
Simpson's Rule, **22**, 69–72, **70**, 72, 75–7, **100**, 101, 143, 149, 166
single chine, 52, **52**
single engine, 43, 112–13
single-screw vessels, 104, 111
skeg, 66, **66**, 67
skew (of propeller), **105**, 106

skin of the hull, 8, 69, **69**, 74
skin resistance, 81, **81**, 84–5
slip, 104, **105**
sloop rig, 37, **38**, 39
sluggishness, 153
space requirements, 119–23, **120**, **123**
spars, 35, 48
specification, 24, 51–5, 56, 88
speed of craft, 16, 63, 78, 94, **105**, 107–9, 112–13
  and resultant wave formation, 82–4, **83**, **84**
Speed:Length Ratio, 16, 78, 84–6, **85**, **90**, 91–3, 92, 107–9
spinnakers, 36, 39
spoon bow, 63, **65**
square rig, 39
stabilisers, **198**, 199
stability, 76, 77, 91, 153–62, 191–3
stanchion, **12**
standing rigging, 35
stations, 9, 117, 136
stays, 35, 36, **36**
staysails, 39
steel construction, 13, 32, **180**
steering, 48, 54, 127–33
stern shape, 63, 64, **65**
stern wave system, 83, 84, **84**
stiffeners, **12**, 13
stiffness, 153
straight bow, 64, **65**
streamline flow, 80, 81, **81**
strength of yacht, 60, **60**, 61, **69**, 75, 163–70
stresses, 73, 165–6, **167**
stringers, **12**
structural aspects of design, 11–13, **12**, **60**, 61, 165–6, 175–7, **177**, **178**
superstructure, 67, **67**
Superstructure Drawing, 172
sweeps, 19, 21, **21**, 137, 139

T

Table of Offsets, 140
tack (of sail), **37**
teak timber, 30, 33, 34, 132, 183
tee-squares, 18, **19**
templates, **20**, 22, 139
tensile stresses, 165, **165**
terminology, 7–16
testing of ship models, 85–7, **86**, 190, **191**
thermoplastics and thermosets, 30
thrust, 104
thrust horsepower (thp), 107
timber, **4**, 30, 32–4, **32**, **33**, **34**, 54, 132, **180**, 183
topping lifts, 35, **36**
topsail, **38**
Total Weight Curve, 166, **167**
transitional flow, 81, **81**
transom stern, 64, **65**
transverse stability, 154, **155**, 156, **160**
transverse waves, 82, **82**
triangular divergent waves, 82, **82**

203

Index

trigonometry, 78–9, **78**, **99**
trim, 11, 161–2, 195
trochoidal waves, 83–4, **197**, 198
true direction of wind, 100
trysail, 39
tumblehome, 10, **11**
turbulent flow, 81, **81**
turnbuckles, 48, **49**
turning force, 127
twin engines, 44, 112–13
twin-screw vessels, 104–5, 111
two-stroke engines, 41
types of hull, 52–3, **53**, **58**, 108–9
types of propeller, **110**
types of yacht, 25–9, 111–13

U

units of measurement, 79

V

ventilation, 49, **50**, 123, 181–2
Vertical Prismatic Coefficient, 14, **15**, 16
vibration, 198–9

volume, 72, 73
   of displacement, 14, **15**, 16, 91, 143

W

wake of vessel, 104, **105**, 111
waterline length (LWL), 8, 9, **9**, 14, **15**, 16, 63, 84, 85, 91, 93, 94, 95, 108, 117, 136, 149
waterlines, 9, **10**, 137
waterplane area, 14, **15**, 16, **91**
Waterplane Coefficient, 14, **15**, 16, 90, 91, 94
wave formation, 82–4, **82**, **83**, **84**
wavemaking resistance, 84, 87
weather helm/rudder, 101
weight, 73, 147–52, **148**, 154, 165–6
weights, 18, 19, **20**, **99**, 137, **139**
wheelhouse arrangements, 123–4, **124**
wind calculations, 98–102, 195
wiring system, 181, 182
wooden craft *see* timber

Y

yard (of sail), **37**
yawl rig, **38**, 39